SMART GREEN WORLD?

In this book, Steffen Lange and Tilman Santarius investigate how digitalization influences environmental and social sustainability. The information revolution is currently changing the daily lives of billions of people worldwide. At the same time, the current economic model and consumerist lifestyle needs to be radically transformed if society is to overcome the challenges humanity is facing on a finite planet. Can the much-discussed disruption potential of digitalization be harnessed for this purpose?

Smart Green World provides guiding principles for a sustainable digital society and develops numerous hands-on proposals how digitalization can be shaped to become a driving force for social transformation. For instance, the authors explain why more digitalization is needed to realize the transition towards 100% renewable energy and show how this can be achieved without sacrificing privacy. They analyse that the information revolution can transform consumption patterns, mobility habits and industry structures – instead of fostering the consumption of unneeded stuff due to personalized commercials and the acceleration of life. The authors reveal how *Artificial Intelligence and the Industrial Internet of Things* pose novel environmental challenges and contribute to a polarization of income; but they also demonstrate how the internet can be restored to its status as a commons, with users taking priority and society at large, reaping the benefits of technological change in a most democratic way.

Providing a comprehensive and practical assessment of both social and environmental opportunities and challenges of digitalization, *Smart Green World? Making Digitalization Work for Sustainability* will be of great interest to all those studying the complex interrelationship of the twenty-first-century megatrends of digitalization and decarbonization.

Steffen Lange is a Postdoctoral researcher at the Institute for Ecological Economy Research, Germany.

Tilman Santarius is a Professor at the Technical University of Berlin and the Einstein Centre Digital Future, as well as Senior Fellow at the Institute for Ecological Economy Research (IÖW), Germany.

Routledge Studies in Sustainability

For more information on this series, please visit www.routledge.com/Routledge-Studies-in-Sustainability/book-series/RSSTY

SMART GREEN WORLD?

Making Digitalization Work for Sustainability

Steffen Lange and Tilman Santarius

Routledge
Taylor & Francis Group
LONDON AND NEW YORK

earthscan
from Routledge

HEINRICH BÖLL STIFTUNG
The Green Political Foundation

First published 2020
by Routledge
2 Park Square, Milton Park, Abingdon, Oxon OX14 4RN

and by Routledge
52 Vanderbilt Avenue, New York, NY 10017

Routledge is an imprint of the Taylor & Francis Group, an informa business

Published in collaboration with the Heinrich Böll Foundation.

British Library Cataloguing-in-Publication Data
A catalogue record for this book is available from the British Library

Library of Congress Cataloging-in-Publication Data
Names: Lange, Steffen, author. | Santarius, Tilman, author.
Title: Smart green world? : making digitalization work for sustainability / Steffen Langen and Tilman Santarius.
Description: Milton Park, Abingdon, Oxon; New York, NY: Routledge, 2020. | Includes bibliographical references and index.
Identifiers: LCCN 2019058425 | ISBN 9780367467579 (paperback) | ISBN 9780367467616 (hardback) | ISBN 9781003030881 (ebook)
Subjects: LCSH: Sustainable development. | Social systems—Growth. | Technology—Social aspects.
Classification: LCC HC79.E5 L3545 2020 | DDC 338.9/270285—dc23
LC record available at https://lccn.loc.gov/2019058425.

ISBN: 978-0-367-46761-6 (hbk)
ISBN: 978-0-367-46757-9 (pbk)
ISBN: 978-1-003-03088-1 (ebk)

Typeset in Bembo
by Wearset Ltd, Boldon, Tyne and Wear
Printed and bound by CPI Group (UK) Ltd, Croydon, CR0 4YY

We wish to express our heartfelt gratitude for the contributions, comments and criticism received from: Jens Bergener, Manuel Brümmer, Vivian Frick, Swantje Gährs, Maike Gossen, Nina Güldenpenning, Richard Harnisch (copy-editing), Bernd Hirschl, Anja Höfner, Jan Osenberg, Johanna Pohl, Nina Prehm, Rainer Rehak, Markus Reuter, Wolfgang Sachs, Benjamin Stephan, Christian Uhle, Norbert Reuter, Laura Theuer, Katharina van Treeck, Oliver Wolff – and, last but not least, to Christopher Hay for translating the book into English.

CONTENTS

ILLUSTRATIONS

Figures

Table

PREFACE

Heinrich Böll Foundation

The development and diffusion of manifold digital technologies and applications is changing societies and economies the world over, often at great speed. Public and political debates have seen many facets fading in and out of the discourse: the prospects of the "New Economy" and the impacts of the "dotcom breakdown", the opportunities of social media for democratization, the perils of filter bubbles and echo chambers, and more recent debates about surveillance, IT security and the manipulation of elections through platform politics. Always high on the political agenda – unsurprisingly – are economic issues, such as the potential of artificial intelligence for economic growth or the risks of automation for jobs.

However, rarely has anyone asked: what does the mega-trend of digitalization imply for global ecology and equity? How do the countless apps, the platforms and the internet influence energy and resource consumption and greenhouse gas emissions? How do big data, the internet of things and robotization impact on income distribution and economic justice? This book is the first of its kind, interlinking the two debates on the role of information and communication technology (ICT) and digitalization on the one hand, and the importance of sustainable development on the other.

But, in doing so, the book goes far beyond the older debate on "Green ICT". No doubt, problems associated with the extraction of scarce metals or toxic materials in the production process of ICT hardware are important aspects – and are indeed included here. In two respects, however, this volume broadens the scope: first, the book not only looks at the production and use phases and end-of-life challenges of hardware. It dives deep into the analysis of the social change associated with digitalization and the prospects of that social change for alternative patterns of production and consumption. Just to give an example: people are not only using the smartphone for navigation in order to travel from one place to another as they always used to do. Instead, smartphones and internet offer new possibilities for changing existing transport structures and habits. Will current ways of digitalizing the transport system deliver a more sustainable mobility – or, rather, boost unsustainable travel modes? Which specific tools best foster a sound mobility transition in cities and rural areas? This book asks these questions for several sectors (energy, consumption, industry, work) that require deep sustainability transition. Second, the book does not restrict analysis to environmental issues. The authors analyse the social and environmental effects of digitalization as interconnected challenges, just as the global concept of sustainable development envisages. For instance, a device that reduces carbon dioxide emissions will only be sustainable if it does not restrict people's prospects for a better life. Likewise, the authors argue that when collaborative platforms facilitate more decentralized and democratic economic production, this should not come at the price of increased energy and resource intensity. The bottom line of this book is: digitalization will only be sustainable if it serves both social *and* ecological improvement.

The authors are excellently equipped to tackle such a comprehensive analysis. Steffen Lange and Tilman Santarius are trained in the disciplines of economics, sociology, politics and anthropology, and have long records in applied sustainability research. Amongst other things, they have worked on climate and energy policy, world trade issues and sustainable economics. Moreover, they can draw on their "Digitalization and Sustainability" interdisciplinary research team at the Technical University of Berlin and the Institute for Ecological Economy Research. The team includes PhD candidates in engineering, psychology, marketing and social science, and spans empirical and theoretical research both at the micro- and at the macro-level. Drawing on this expertise, Steffen Lange and Tilman Santarius distil key insights to answer the big question: to what extent can digitalization contribute to sustainable development? Readers may find it particularly valuable that the authors provide copious references to publications that address not just scientific audiences but are easily readable for the interested public. This book seeks to spur public and political discourse and is hence designed to be an easy-to-read compendium that widens the horizon and the focus of the debate.

Most notably, this book is not concerned only with pointing the finger at critical aspects of digitalization, such as energy and resource consumption, which have so far been largely neglected. To the contrary, it is an optimistic book that develops proposals guided by the question: how can society and politics actually shape technology development and application so that it contributes to the greater good? The book presents policies and measures as well as regulations that generate beneficial settings in which sustainable digital technologies can flourish. It outlines what users can do in their everyday lives. And it discusses which role civil society organizations can play – in agenda setting, digital education and political lobbying. Hence, the book combines a profound systemic analysis of opportunities and risks with transformative knowledge of how to realize the potential of digitalization to contribute to an ecological and just transition towards sustainability.

The Heinrich Böll Foundation has long been dedicated to the search for solutions that spur global sustainable development. With this book, we hope to kick off a new debate in North and South on the role of technology for the common good.

Sincerely,
Barbara Unmüßig
President, Heinrich Böll Foundation

ACKNOWLEDGEMENTS

This publication is based on research in the project "Digitalization and Sustainability", which is funded by the German Federal Ministry of Education and Research as part of its "Research for Sustainable Development Framework Program"/"Social-Ecological Research", Funding no. 01UU1607A.

The translation of this book from German into English was kindly financed by the Heinrich Böll Foundation, Germany.

1
DISRUPTION FOR A SUSTAINABLE FUTURE?

"Everything will change. It will be a very different world!" This prophetic message from the IT industry has now taken hold as the popular view of how digitalization will impact on society. Countless newspaper and blog articles on digitalization start with the comment that it will transform our lives. But that's not all: many believe that digital innovations are "disruptive", shaking up and revolutionizing business, communication, manufacturing processes and consumption habits.[1] Even critics often leave no room for doubt that we are in the throes of a digital revolution.

And, indeed, never before has any technology taken hold so swiftly or had such a profound impact on our daily lives. In less than ten years, the tiny devices that make information about everything and nothing available any time, any place, have become the constant companions of much of the world's population. Yet the "smartphonization" of our lives will not be the end of this process. The internet of things, big data, artificial intelligence, smart cities and virtual reality all feed into today's vision of a future world with potentially profound implications for many areas of life and the economy. We do not yet know how much of it will become reality. But we should prepare ourselves for a future marked by digitalization. Only one question remains unanswered: will it change our society for the better?

The developers of digital technologies are by no means the first to aspire to change society, nor are they alone in this ambition. Many sustainability researchers and representatives of civil society have been insisting for years that our economic model and consumerist lifestyle need to change. They are worried that our Blue Planet's carrying capacity could soon reach a tipping point, putting entire communities at risk of destabilization.[2] And there is no shortage of worrying headlines. Climate change is advancing relentlessly, with potentially devastating consequences for biodiversity loss and human communities over the next few decades.[3] At the same time, growing numbers of people – not only in the world's poorest regions, but increasingly also in the heartlands of the early industrialized countries – are struggling to achieve job and income security and a decent life in society.[4] Here, too, we should prepare ourselves: without a radical restructuring of our economy and lifestyle, our future is likely to be dominated by environmental and social crises.

Like many others, we – the authors – are convinced that a fundamental transformation is essential to future-proof our societies.[5] This means that the way in which we produce and consume must be radically overhauled to become sustainable and equitable. In other words, we need major changes if we are to overcome the challenges facing our 21st-century world, but with one clear goal: these changes should advance a social and ecological transition. Can the much-discussed disruption potential of digitalization be harnessed for this purpose?

Our generation faces two Herculean tasks. We must build a fairer world for a population soon approaching nine billion people and, at the same time, save the environment from collapse. Justice and ecology are the two priorities and are intimately connected. If inequality increases, with dwindling numbers of people having the prospect of a decent life, so the willingness and the capacity – including the financial capacity – to invest in restructuring our economy and society and experiment with more sustainable forms of production, consumption, mobility and housing will dwindle. And, if climate change, the erosion of fertile soils, species extinction and the overexploitation of finite resources deprive our children and grandchildren of the vital bases of life and economic activity, social conflicts will mount, with more people within rich and poor countries around the world facing job losses, social exclusion and poverty.

The heart of the matter is this: without justice, there can be no protection of the environment and, unless the environment is protected, there can be no social justice.[6] That is why greenhouse gas emissions from fossil fuels (coal, oil and gas) in most developed countries must be reduced to zero(!) in the next ten to 15 years as their equitable contribution to global climate action, to keep global warming under the dangerous 2°C threshold and, if possible, to limit it to 1.5°C.[7] Within the next two decades, natural resource consumption must decrease to 10 per cent of its current level in order to safeguard the regenerative capacity of ecosystems and the biosphere.[8] At the same time, the gap between rich and poor – between the highest and the lowest income groups within society – must not be allowed to widen. On the contrary, it must close in order to safeguard peaceful social relations and democracy in the long term.[9] And, in parallel, every individual must have the income needed to participate in society – be it through a job that pays a decent wage or through another source of income.

All these goals are still out of our reach. Progress in most (developed) countries is almost imperceptible. Yes, there is a growing recognition that environmental and social policies are important and that simplistic solutions, such as the firmly-held notion that economic growth is the magic bullet, will not work.[10] However, there is little prospect, at present, of a major social and ecological transition. Most companies are still prioritizing growth over radical change, and even those that lead the field in sustainability have only limited capacity to escape systemic constraints. The majority of people are still caught up in their consumer habits and often willingly embrace new opportunities for consumerism. Most politicians seem to be resorting to the safe option of equivocation – while the influence is growing of populists who cling to an outdated status quo while simultaneously undermining our democracy. There appears to be little prospect of any fresh ideas, social or environmental, from our politicians. Yet, if the culture of public discourse is toxic, the functionality of our democratic institutions is undermined and wars and conflict increase around the world, then a peaceful transformation of society towards more sustainability will become even more of an uphill task.

So, we face a real mega-challenge – to transform our society and make it more sustainable – just as the mega-trend of digitalization is exploding into our lives. Can the disruptive potential of digitalization help to initiate the change that is urgently needed and make tomorrow's world more sustainable and equitable? Certainly, there is no shortage of ambitious statements of intent from the IT industry and Silicon Valley.[11] Facebook founder Mark Zuckerberg has said that he wants to build a global community that works for all of us. Elon Musk aims to popularize the Tesla, an eco-friendly car. And Microsoft founder Bill Gates' ambition is to end poverty and hunger in the world. But, is the rhetoric of these and other companies matched by action? How have digital information and communication technologies, the internet, the countless apps and digital platforms influenced energy and resource consumption, jobs and income distribution thus far? And how will increasingly interconnected and rapid communication by people, things and machines affect the environment and fairness in future?

In this book, we look for answers to the core questions that are likely to determine the future of humankind. How can digitalization help to preserve the biosphere and increase social justice? What opportunities and risks are presented by the increasing digitalization to transform our lifestyles and economic structures towards greater sustainability?

Notes

1 See e.g. Harari, *21 Lessons for the 21st Century*, 2018; McQuivey, *Digital disruption*, 2013.
2 See e.g. Hirsch, *Social limits to growth*, 1995; Jackson, *Prosperity without Growth*, 2011; for a very early discussion see Meadows *et al.*, Limits to growth, 1972; and many others.
3 IPCC, *Climate change 2014*, 2014; IPCC, *Global warming of 1.5°C*, 2018; Steffen *et al.*, Planetary boundaries, 2015.
4 See e.g. OECD, *In It Together*, 2015; O'Neill *et al.*, A good life for all within planetary boundaries, 2018; Raworth, A safe and just space for humanity, 2012.
5 See e.g. Brand/Wissen, Global Environmental Politics and the Imperial Mode of Living, 2012; New Economics Foundation, The Great Transition, 2009; Polanyi, *The great*

transformation, 2010[1944]; German Advisory Council on Global Change (WBGU), *World in Transition A Social Contract for Sustainability*, 2011.

6 Sachs, *Planet dialectics*, 2015.

7 The New Climate Institute, What does the Paris Agreement mean for climate protection in Germany?, 2016; UNFCCC, Paris Agreement, 2015; see also IPCC, *Global warming of 1.5°C*, 2018.

8 Meadows *et al.*, *The limits to growth*, 2004; Sachs/Santarius, *Fair future*, 2007; Weizsäcker *et al.*, *Factor 5*, 2009.

9 Piketty, *Capital in the Twenty-First Century*, 2014; Wilkinson/Pickett, *The Spirit Level*, 2010.

10 Heinberg, *The end of growth*, 2011; Jackson, *Prosperity without Growth*, 2011; Lange, *Macroeconomics Without Growth*, 2018; Victor, *Managing without growth*, 2019.

11 See Morozov, *To save everything, click here*, 2013.

2
HIDDEN DRIVERS OF DIGITALIZATION

It is a commonplace, in discussions about digitalization, that technology is "neutral". Technology, it is claimed, is neither good nor bad, but is simply a value-free tool that can be utilized in pursuit of a diverse array of goals. According to this line of argument, a technology makes no judgements about the purpose for which it is used. A motor vehicle, for example, can be used to transport the sick to hospital, be pressed into service as an armoured car in wartime or taken out for pleasant weekend excursions. However, this interpretation applies, at most, to technology in an abstract sense; in other words, to the purely theoretical notion of an "automobile" – or of "digitalization". But, in its tangible form, technology always embodies the interests and intentions of its makers. The specific form that technology takes is determined by those who design, manufacture and distribute it, and regulate its conditions of use. What would cars look like if they were not designed to meet the interests of the automotive industry, oil companies, petrol station operators, road construction firms, transport policy-makers and, of course, the motorists' lobby? Indeed, would they exist as private vehicles, or would we be reliant on public transport?

It is no different for digitalization, which we broadly define as the permeation of various ICT devices and applications (hard- and software) into diverse areas of everyday life, society and economy. What can they be used for – and what not? Which needs are fulfilled – and which are encroached upon? How each individual digital device is designed and each application programmed, how search engines provide information and the internet is regulated will never be "neutral". Determining who shares the benefits of digitalization is not just a matter for users alone. Technological development is the result of an ongoing process of social dialogue and conflict, and the outcomes are often determined by the power relations of different stakeholders involved.

To enhance our understanding of which issues are of particular concern in the context of digitalization, it is helpful to cast a glance at its history. And this is revealing, for it shows that those who initiated the development of computers and the internet still influence the design, regulation and use of digitalization in its highly diverse forms today. Of course, over the decades, countless scientists, engineers and practitioners had a hand in making information and communication technologies and the internet what they are today. But, if we look back at the early days, three interest groups stand out as the godfathers of digitalization: the military, the industry and the counterculture of those working for a better world.

The US military can be credited with the first attempts to develop and network digital information and communication technologies. Communications are not only extremely important in modern warfare; they also play a key role in counter-espionage against potential enemies and in the development of automatic weapon systems. In the wake of the Sputnik crisis in 1957, when the Soviet Union became the first nation to go into space, thereby teaching the West the meaning of the word "fear", the US Department of Defense set up the Advanced Research Projects Agency (ARPA) and tasked it with developing a flexible, autonomous computer communications system without a central core that would be as robust as possible in the event of a nuclear war. A good ten years later, in 1969, ARPANET came into operation. One of the most important precursors of the internet, it was used for military applications from 1975 onwards.[1] Of course, in later years, numerous scientists who were not directly subordinate to the Department of Defense were also involved in further developing ARPANET. However, most of the funding for the computer sciences from the 1950s until the 1970s came from the military budget.[2] Therefore, even researchers and engineers who may well have been working purely for the furtherance of knowledge or in pursuit of the noble dream of using computer-based communication to change the world ultimately owed their contracts and resources to the Cold War, in which the US-led NATO countries aimed not only to win the arms race but also to out-smart the Soviet Union and its allies in the counter-espionage stakes.[3]

Industry came relatively late to information and communication technologies – but then wasted no time in shaping them for its own purposes. In the 1950s, most computers in existence were operated by the military and academic institutions, but, from the 1960s onwards, industry surged to the fore. From then on, it was mainly banks, insurance companies and, increasingly, the major industries that controlled the bulk of the processing capacity.[4] In 1969, the Programmable Logic Controller (PLC) was invented, enabling manufacturing processes to be programmed and guided by computer. The leaps in industrial automation in the following decades enabled companies to replace much of their workforce with machines, expand production and increase their profits. From the 1970s onwards, private industry thus became a key driver of digitalization, with telecoms companies leading the way. AT&T, an American multinational conglomerate, is one example: by exerting considerable political pressure, it secured the sweeping deregulation of the telecoms market, thus ensuring that computer data is able to use the same networks as telephone data – with minimal regulation. Hard- and software companies such as IBM, Intel, Microsoft, Oracle and others were also instrumental in driving this development and soon became some of the world's most financially powerful corporations, with industry chiefs such as Bill Gates and Larry Ellison long counting among the world's wealthiest billionaires.[5] In these transnational corporations, systems designers established the bases for the rollout of PCs and other digital technologies and for networking via the internet, long before Tim Berners-Lee invented the World Wide Web in 1991. Many companies also operated their own intranets. By the end of the 1980s, for example, the Citicorp bank operated what was then the world's largest private intranet, handling forex transactions totalling around US$200 billion per day from its 94 national locations.[6] Digitalization thus laid the groundwork for the transcontinental expansion of the corporations' production and supply chains. And, thanks to digitalization, the value of knowledge and information was captured systematically, evolving steadily into the 21st century's most lucrative business.[7]

Somewhat paradoxically, given the military's interest in monitoring and surveillance and industry's profit motive, early digitalization was also influenced by the alternative movement in the 1960s and the 1970s, particularly American hippy and counterculture.[8] The Free Speech Movement – a precursor of the student revolution – was a large-scale protest for justice and freedom of expression in 1964. With the so-called "military-industrial complex" as one of its targets, the movement campaigned for a society in which technology was placed back in the hands of ordinary people, with workers no longer cogs in the industrial machine.[9] In the late 1960s and early 1970s, a grassroots movement in the USA and worldwide began seeking alternative ways of living in harmony with nature, liberated from capitalism and its curbs on freedom. This movement was not necessarily hostile to technology: many types of small-scale "convivial technologies"[10] for a variety of uses, including information-gathering, were seen as essential tools for achieving liberation and independence from industrial capitalism. The 1968 issue of the *Whole Earth Catalog* – which provided information and tools to support communities towards self-sufficiency – is regarded as an important analogue precursor of the internet, for example.[11] The values and ideals of the counterculture influenced the "tech community" from the very beginning. Numerous IT companies were set up by long-haired hippies in backyards, Steve Jobs founded Apple as a "countercultural computer company" and an entire generation of hackers emerged, operating according to their own strict code of conduct (the "hacker ethic").[12] And, while some pursued more individualistic or libertarian goals, many hackers saw – and still see – digitalization as an opportunity to dismantle the hierarchies of oppression, to paralyse exploitative corporations and to replace destructive capitalism with an economy built around the environment and social justice.[13] Unsurprisingly, attempts at counter-espionage by the military and intelligence services, but also the digital titans – especially Microsoft, were regarded by the digital counterculture as the enemy, as rogue initiatives, and were targeted from the outset.

These three highly diverse groups – the military, industry and counterculture – thus drove the development of digital technologies in the early days. But, which of them left the strongest mark on digitalization today? Is digitalization mainly about surveillance and monitoring? Are efficiency and higher profits for global capitalism its aims? Or is its purpose to promote self-determination, social coop-eration and a sustainable economy? These are the three major issues that con-tinue to dominate the discourse in every fresh wave of digitalization. Sometimes it is difficult to figure out whose interests prevail, as the following three topical examples show.

Let's look first at smart home systems. These systems, many of which are cur-rently in development, enable homeowners to control services such as heating remotely using smartphone technology. This may save energy, which would be good for the climate. However, many of these systems are built around sensors in the home. The connected devices produce detailed profiles of the user's move-ments and gather information about their intimate behaviour to an extent rem-iniscent of the level of surveillance offered by an offender's electronic ankle tag. Are these smart home systems really about energy saving? Are they a manifestation of the security industry's obsession with surveillance? Or do they serve companies' interest in selling, not only new electronic devices, but also users' personal data?

Let's move on to the second example. Thanks to artificial intelligence, robots and personal digital assistants will soon replace the human workforce in many areas of activity that are not yet fully automated. Will this emancipate people from onerous labour, erase traditional class divisions and improve everyone's opportunities to engage in self-determined and creative work? Or will it lead to mass redundancies and the redistribution of income away from the newly unem-ployed in favour of the robots' owners and developers?

Our third example is blockchain technology, which – according to alternative futurists[14] – has the potential to replace all forms of intermediary, even includ-ing (central) banks and trading companies, in a radically democratized economy. Yet, surprisingly, it is the international banks and other financial institutions that are investing heavily in the rollout of Bitcoin and other blockchain applications. Will this technology genuinely revolutionize capitalism? Or will it simply make it more efficient and further accelerate the global movement of capital?

We will be exploring these and many other questions in this book. The answers will depend on which players develop, design and utilize the various digital technologies and applications, and their motives for doing so. Far from being "neutral", these technologies are often inherently contradictory. Over and again, the fundamental question arises: what is the primary purpose of digitaliza-tion? Is it to support monitoring and surveillance? Is it about commercialization?

Or does it serve the transformation towards a more equitable and sustainable society?

There are, of course, passionate advocates and fervent critics of each of these three purposes. Some still regard digitalization as a milestone in the struggle for freedom of speech and real democracy.[15] Others fear that it may lead straight to a digital dictatorship that crushes our privacy and democratic rights.[16] Many countries' governments, including China and Germany, regard the industrial internet of things as the best way to stimulate economic growth, safeguard jobs and boost incomes.[17] However, many critical researchers are warning that a significant share of humanity faces the prospect of joblessness or precarious employment within the next few decades.[18] Some believe that digitalization is a powerful tool for protecting the climate and resources and building a cooperative social economy.[19] Others, in turn, fear that digitalization may well reinforce capitalism's ruthless exploitation of our environment by facilitating "smart dominion" over nature.[20]

This diverse array of opinions, hopes and fears oscillates between hype and hysteria. Critics paint apocalyptic scenarios, placing the blame for every conceivable evil firmly at the door of digitalization. Utopian technophiles, by contrast, rush to proclaim the benefits of the 'next big thing' in the digital revolution long before it is clear whether this "thing" is likely to take hold and whether anyone wants it anyway. Indeed, one might be forgiven for thinking sometimes that the real purpose of these shock-horror reports is to pave the way for ultimate blind acceptance of the next wave of technology in our lives.[21] Perhaps nothing will be as bad – or as beneficial – as is claimed. On the contrary, perhaps it will simply serve to perpetuate the – sadly quite unsustainable – status quo.

Ultimately, all the hopes and fears may – or may not – prove to be true. That will depend on what our society makes of digitalization and which manifestations of digital technology we as individuals wish to embrace and utilize. In answering the question whether digitalization is more likely to lead us in the direction of surveillance, commercialization or sustainability, what matters most is the social and political course that is set. Neither politicians nor researchers or civil society have begun to frame an agenda for a transformative digital policy. We believe it is long overdue.

In the following chapters, we start by analysing the opportunities and risks that digitalization presents for the environment: for reducing energy and resource consumption, cutting greenhouse gas emissions and supporting a green transformation of production and consumption (Chapter 3). We then consider how digitalization can influence the various dimensions of economic justice – jobs, concentration of power, income justice and economic growth (Chapter 4). Based on the findings of our analyses, we then develop three guiding principles for sustainable and equitable digitalization (Chapter 5). And, finally, we look at their implications for politics, developers and users of technology, and civil society (Chapter 6). Which frameworks are needed and which policies and measures have the capability to steer and support digitalization? Which consumption and behavioural patterns are facilitated by such settings and policies? And how can critical civil society help to ensure that digitalization genuinely contributes to social and ecological transformation?

Notes

1 Castells, *The Rise of the Network Society, The Information Age*, 1996.
2 Castells, *The Internet galaxy*, 2001, p. 18.
3 See also Leslie, *The Cold War and American science*, 1993.
4 Schiller, Digital capitalism, 2000; ZUBOFF, *The age of surveillance capitalism*, 2019.
5 O'Regan, *A Brief History of Computing*, 2012.
6 Schiller, *Digital capitalism*, 2000, p. 13.
7 See e.g. Mason, *PostCapitalism*, 2015; Rushkoff, *Throwing rocks at the Google bus*, 2016.
8 Turner, *From counterculture to cyberculture*, 2010.
9 Cohen/Zelnik, *The Free Speech Movement*, 2002.
10 Illich, Tools for conviviality, 1973.
11 Akos, *Whole Earth Catalog* (1975), 2014; Turner, Where the counterculture met the new economy, 2005.
12 Levy, *Hackers*, 2010; Rothstein, A Crunchy-Granola Path From Macramé and LSD to Wikipedia and Google, 2006.
13 Rifkin, *The zero marginal cost society*, 2014.
14 See e.g. Tapscott/Tapscott, *Blockchain revolution*, 2016; Voshmgir, Blockchains, Smart Contracts und das Dezentrale Web, 2016.
15 See in Winkel, The Perspectives of Democratic Decision Making in the Information Society, 2016.
16 See e.g. Helbing *et al.*, Will Democracy Survive Big Data and Artificial Intelligence?, 2017; Welzer, Smart Dictatorship, 2016.
17 See e.g. Bundesministerium für Wirtschaft und Energie, Industrie 4.0, 2019.
18 See e.g. Bartmann, *The Return of the Servant*, 2016; Dämon, Studie Digitalisierung und Arbeitsplätze, 2015; Frey/Osborne, The future of employment, 2013.
19 See e.g. GeSI/Deloitte, Digital with Purpose, 2019; Rifkin, *The zero marginal cost society*, 2014; Sachs *et al.*, ICT & SDGs, 2015.
20 See e.g. Kopp *et al.*, *At the Expense of Others?*, 2019; Pilgrim *et al.*, The Dark Side of Digitalization, 2017.
21 See e.g. Helbing *et al.*, Will Democracy Survive Big Data and Artificial Intelligence?, 2017.

3

SAVING NATURE WITH BITS AND BYTES?

The biggest library in the world is the British Library in London. The fact that the most comprehensive collection of books, magazines, sound recordings, maps, prints and drawings, and many other publications, has been maintained in England for so long demonstrates the hold that the tradition of preserving physical sources of information has over us. The library's history dates back to 1753, the time of the British Empire. The collection now comprises some 200 million items, including 25 million books, housed in a building that – with its floor area of 111,500 square metres – is one of the largest in the UK. If all the library's books were to be saved as e-books, they would amount to an estimated 65 terabytes of data. Given that a 3.5″ hard disk can hold 12 terabytes of information, the contents of the 625 kilometres of shelving in the British Library could be contained in three external hard drives on a small side table next to our laptop. At first glance, it seems obvious that digitalization could potentially save resources on a vast scale.

Accordingly, in many quarters, great hopes are being pinned on digitalization as the answer to environmental problems. For example, it is assumed that a digitalization of factories and industries will make the economy more sustainable, because this can make a significant contribution to resource conservation and energy efficiency.[1] This expectation is based on the assumption that digitalization will enormously enhance the productivity of natural resources and energy.[2] According to studies conducted by management consultants and the Global e-Sustainability Initiative, which brings together major telecommunications and IT companies from around the world, information and communication technologies could cut worldwide CO_2 emissions by an astonishing 20 per cent by 2030.[3] If this is the case, digitalization would make a significantly larger contribution than the sum total of all climate legislation to date.

However, it is debatable whether all the study's assumptions are plausible; we shall return to this point shortly. While digitalization presents major opportunities in areas such as public transport and the transition to sustainable energy systems, in other areas – such as consumption and industry – it is questionable whether digitalization can really contribute to environmental sustainability. In this chapter, we explore the opportunities and risks of digitalization in connection with reducing energy and resource consumption and making production and consumption more sustainable. We shall examine five areas:

First, we look at the material basis of all the devices and infrastructure components – from the smartphone to the server park – that underpin virtual and apparently immaterial digital applications. Can materials be saved and greenhouse gas emissions reduced by using substitutes for physical products – for example, by reading books on an e-reader or streaming films instead of borrowing or buying a paperback or DVD? We will show that comparing the environmental footprint of digital services with that of conventional, "analogue" ones often reveals no significant differences. It is, however, interesting to consider how the outcome changes when rebound effects are taken into account.

Second, we consider how digitalization can in future help us generate electricity and heat without needing to use coal, oil and gas and how it can ensure a stable renewables-based supply. It becomes apparent that the biggest environmental and social opportunities of a decentralized, democratic energy system cannot be unlocked without the help of digitalization. One of the key challenges here involves taking appropriate steps to protect the private sphere.

The third major area that we look at is our consumption. We highlight ways of adopting sustainable consumption patterns, for example through sharing, do-it-yourself or avoiding new purchases. We also show how digitalization boosts already high levels of consumption through personalized advertising and omnipresent shopping opportunities.

Alongside rising consumption, we are seeing, fourth, an increase in transport that is itself only made possible by digitalization and that could swell enormously with the advent of autonomous (private) vehicles. But, we also show how – if properly managed – digitalization of vehicle sharing and public transport could have great potential for environmentally sound transformation of the transport system.

Fifth, we look at production, where digitalization underpins the concepts of an industrial internet of things and of 'Industry 4.0'. We describe the opportunities for boosting material and energy efficiency through digitalization, but also show how new growth could lead to rebound effects and thus wipe out environmental benefits.

Of course, all these facets of digitalization have implications in addition to their environmental impacts. The developments we describe also affect which jobs are created and lost, how incomes are (re)distributed and who gains and who loses economically. We address these issues in Chapter 4. So, let us now look at the environmental impacts of digitalization. We start with the technical devices that we can hardly imagine being without: smartphones, PCs, tablets and the like.

The physical basis of the virtual world

If asked to consider digitalization from an environmental point of view, you might think first of the electricity consumed by all the devices that are now part of our lives. A few years ago, there was discussion in the media of whether a single Google search, using four watts per hour, might consume as much electricity as an energy-saving light bulb left on for an hour; Google itself explained that a search uses "only" six minutes' worth (about 0.4 watt-hours).[4] Other people may wonder how much electricity the computer and the data transfer process are using when they stream a film. We shall look at examples of this sort below, because they are representative of the current state of statistics and research on the subject. But, when adopting this approach, it is easy to forget that digital devices, digital infrastructure and applications also consume resources and energy during manufacture. And, finally, devices must be disposed of as electrical and electronic waste at the end of their – often short – lives.[5]

We shall examine the material basis by looking at the smartphone. Since the first iPhone came out in 2007, the smartphone has become the must-have accessory – at least for wealthier members of the world population. Each individual phone is small and light, with a sparkling finish. Unlike the key technologies of previous periods of industrialization – smoke-belching locomotives, cumbersome cars and noisy aircraft – the environmental impact of each device seems negligible: making the average smartphone involves just five grams of cobalt, 15 grams of copper, 22 grams of aluminium and several other resources. But, it is the overall quantity that matters. More than seven billion smartphones have been sold around the world in just the first ten years, having eaten up 38,000 tonnes of cobalt, 107,000 tonnes of copper, 157,000 tonnes of aluminium and thousands of tonnes of other materials during that period.

It is true that smartphones use only 1 to 3 per cent of the global primary production of most metallic ores (around 10 per cent in the case of cobalt and palladium).[6] But, smartphones are just one device among many. Digitalization involves the production of all sorts of other devices such as PCs, tablets, mp3 players and other wearables. The digital infrastructure must also be set up and operated – we are talking about all the data cables, servers and data centres that we never get to see, but, without which, we could not access the internet. All of this also uses millions of tonnes of resources. For example, 25 per cent of all the silver mined worldwide goes into electronic products. Like silver, other resources used by the electronics industry – such as cobalt, tantalum, platinum and palladium – also come from countries in which people work in appalling conditions, are exposed to major health and safety risks, and are often paid pitiful wages.[7] In addition, the majority of digital devices end up as electronic waste in the poorer countries of the world where health care, occupational safety and economic fairness are low priorities and improper disposal practices may cause environmental pollution. In 2015 alone, 42 megatons of electronic devices were discarded worldwide, and it is predicted that in 2020 this figure will reach 52 megatons.[8] This mountain of electronic waste is roughly equivalent to a scrapheap of all the 46 million cars that are currently on Germany's roads.

Like the smartphone, all digital devices and digital infrastructure work only when the current is flowing. Electricity is needed, too, for their manufacture. Production of the seven billion smartphones made over the last ten years will have used around 250 terawatt hours of electricity – an amount comparable to the total annual electricity consumption of a country such as Sweden or Poland.[9] While most business sectors in the world, either naturally or by way of political regulation, have become more energy efficient, the sector producing ICT hardware, unfortunately, is among the few sectors that have actually increased their energy intensity over the past couple of years.[10]

With regard to the use of the digital end devices, the initial news is good: they are becoming steadily more energy-efficient. But, this energy-saving potential is wiped out by the fact that processors are becoming ever faster and processing power – and screen size – are constantly increasing. A long-term trend has been identified: processing power per kilowatt-hour has doubled every 18 months.[11] But, it is not only processing power that is increasing[12] – the amount of energy

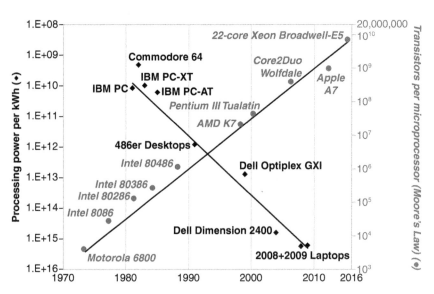

Power consumption for a given level of processing power halves roughly every eighteen months (Koomey's Law). Processing power doubles in roughly the same period (Moore's Law). This results in a rebound effect: efficiency increases are offset by faster and more powerful processors.

FIGURE 3.1 Energy consumption in relation to processing power and processor capacity

Source: Authors' diagram, based on Koomey et al. (2011); Roser and Ritchie (2017).

used by screens and data centres has also risen sharply. Digitalization is thus a textbook example of the rebound effect:[13] technical efficiency increases result in a rise in consumption, negating the potential for savings.

Nevertheless, many users will feel that charging their mobile phone or using their router is of little significance on a global scale. And it is true that, for many digital devices, electricity consumption in the use phase is declining. But, manufacturing the new and ever more powerful devices needs more energy – this is another factor that is counteracting the potential savings in the use phase.[14] In addition, more and more computing capacity and storage space, and an increasing number of services, are located in the cloud. With our smartphones, we are now utilizing the computing power of data centres and thus shifting some of our electricity consumption to the providers of the apps and digital services. The electricity used by ICT now accounts for about 10 per cent of global electricity usage; by 2030, this could rise to 30 or even 50 per cent (see also Figure 3.3 below).[15] If the whole internet were a country, it would rank number three in global electricity demand, just after China and the USA. This huge level of electricity consumption can be illustrated thus: if the 2,500 terawatt-hours of electricity used annually by all ICT devices were to be produced by pedal-powered generators, each producing 70 watts of electric power, it would need all eight billion inhabitants of this planet to pedal in three shifts of eight hours for 365 days per year. The good news: electricity consumption of digital devices is included in these numbers, so the pedallers can use their smartphones while on the job.[16]

The present global electricity mix means that most of the electricity used by information and communication technologies comes from climate-damaging coal-fired power plants or even from nuclear power stations.[17] Although Apple, Google and some other companies are setting a good example and drawing most of the electricity for their servers from renewable sources,[18] the overall proportion of electricity from renewables in the ICT sector remains low. Hence, the more we come to rely in all areas of the economy and in our private lives on digital solutions that depend on electricity, the more important it becomes to ensure that the entire global electricity supply is based on renewables. In the course of the discussion here, we shall also show that the energy system itself must be significantly digitalized if the shift to fully renewables-based electricity generation is to be achieved (see section "Distributed energy systems call for digitalization").

With regard to energy and resource consumption, it has to be conceded that in terms of its materials base and energy use, digitalization is at present far from sustainable – quite apart from the fact that many digital devices have only a short lifespan because they are based on (planned) obsolescence or quickly go out of fashion.[19] This brings us to the question of the environmental impact of digital devices during their use phase. Do they contribute to dematerialization to such an extent that our global consumption of energy and resources is reduced to a sustainable level? And does this remain true even when the energy and resources used in manufacturing the devices are taken into account? This is the question we shall consider next.

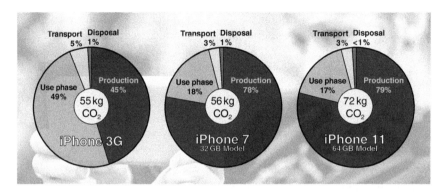

In relation to processing power and screen resolution, the energy efficiency of Apple iPhones has risen significantly. Yet emissions over the entire life-cycle of the product have remained constant over several years, with a diminishing share of energy use attributable to the use phase being countervailed by an increasing share attributable to the hardware production. After much bigger devices have been introduced, the iPhone 11 (2019) now accounts for a 30% higher carbon footprint than the iPhone 7 (2016).

FIGURE 3.2 Carbon footprint of smartphones

Sources: Apple (2009), Apple (2017), Apple (2019).

Do digital devices promote dematerialization?

Despite the ecological footprint of digital devices in the production and use phase, there are great potentials that digitalization can help dematerializing many areas of economic and social life. For example, records and CDs are disappearing from our homes now that we can stream almost all the music in the world. More and more people are no longer reading physical newspapers, but are obtaining their information online. And when did you last use a road atlas or a paper street map rather than an online map service? The expectation that we can reduce our overall resource consumption through digitalization feeds on the haptic experience provided by the miraculous little technical devices that we have at our fingertips: just think of the number of applications we can access with a laptop or even with a little smartphone that fits in a coat pocket. Surely the fact that countless material objects and devices can be replaced by a smartphone, tablet or some other omnipotent handheld device must mean that resources can be saved?

This question may sound simple, but, in fact, it is not easy to answer and requires some complex calculations of environmental impacts. Let us take e-book readers as an example. Advertising images suggest that, by buying a small, slim e-book reader, we remove the need to produce vast numbers of weighty books. The felling of the trees that must be sacrificed to make the paper, the chemicals for the printing ink, the book cover and the binding, the energy and resources that go into logistics and selling, and not least the journey to the bookshop or the delivery to our home by the online seller – all these can be dispensed with.

But, to find out whether the e-reader really delivers environmental benefits, we must compare these aspects of book production with the features of an e-reader in a lifecycle assessment. Producing an electronic device is clearly far more energy- and resource-intensive than printing an individual book. Most e-readers weigh less than 200 grams, but making one uses about 15 kilogrammes of various materials (mainly non-renewable metals and rare earths) plus 300 litres of water, while generating about 170 kilogrammes of carbon dioxide.[20] And it is not just the quantities of inputs and outputs that need to be considered, but also their environmental impacts. There are major differences between e-readers and books, in particular with regard to the toxicity of the materials and manufacturing processes involved. While the paper industry in many countries (still) has highly undesirable environmental impacts – for example, if chlorine and acids pollute local lakes and rivers, the impacts of the electronics industry are not much better either. E-readers and other IT products contain brominated flame retardants, phthalates, beryllium and many other chemical substances that are highly harmful to health and the environment.[21] Often, too, there are social consequences: for example, some of the cobalt, palladium, tantalum and other minerals used in the manufacture of digital devices is mined under atrocious working conditions in dictatorships such as the Republic of Congo and in other countries of the global South, and, at the end of their life, devices are disposed of as environmentally polluting electronic waste.[22]

The e-reader may nevertheless score better than the book.[23] This ultimately depends on two factors: how many books are read on the e-reader in the course of its life, and the number of people who share an "analogue" book. If the high environmental costs of making the e-reader are to be worth paying, a certain number of books must be read on the device. This number is between 30 and 60 – depending on the size of the books and the environmental indicator used.[24] If fewer books are read on the device, it is better to go down the paper route. Once this number of books has been read, each additional e-book is more environmentally friendly than its analogue counterpart. In addition, the way in which books are used is important, even after the e-reader has reached the environmental break-even point so that only energy consumption and CO_2 emissions then need to be taken into account. If someone buys a book and allows no one else to read it, the file on the e-reader is around five times more energy-saving than the book on the shelf.[25] However, the advantage disappears if several people share a book. Whether the book or the e-reader is better is therefore a question that will have different answers in different situations. Overall, though, it is doubtful whether all the e-readers that are sold are on average used sufficiently intensively – before they break or become technically obsolete – for an environmental benefit to be achieved.[26]

Studies have also compared the analogue and digital versions of other products and services, such as print and online media.[27] We shall now consider listening to music and watching films – analyses of media use show that many users spend significantly more time on these activities than they do reading books.[28] Studies highlighting the potential of dematerialization through digitalization in connection with music consumption started to emerge some time ago.[29] We will pick out one environmental indicator, the carbon footprint. Depending on the assumptions made, downloading an mp3 file saves between one and two kilogrammes of carbon dioxide per album by comparison with purchasing a CD – heavily influenced by the means of transport that the buyer of the CD uses to get to the shop.[30] The manufacture of an iPod produces about 70 kilogrammes of CO_2.[31] From the point of view of climate change mitigation, it follows from this that, for more than about 50 album downloads – about 600 songs, it is worth buying an mp3 player. But, is it actually appropriate to compare the emissions balance of the CD player with that of the iPod in such calculations? In principle, of course it is. But, it is also true that attics and cellars everywhere are full of no-longer-wanted record players, cassette recorders, compact stereo systems, Walkmans, ghetto blasters, CD players, Discmans, digital audio tapes, minidisc players and all the others products of various more or less fleeting eras in the history of entertainment electronics – often still in working order. Now mp3 players are to be added to the pile. And, when one takes on board the fact that music consumption has increased significantly over the years – in part because digitalization makes it possible to access music anywhere at any time while also reducing the financial costs to users, then the music industry as a whole can hardly boast that it has resulted in dematerialization.[32]

As we have said, mp3 players are themselves now being relegated to the scrap heap: people are listening to music on their mobile phones and turning to streaming as a delivery method. Streaming – of music and, in particular, videos – now accounts for an astonishing 70 per cent of all data traffic.[33] This includes the legal and semi-legal streaming of cinema and television films via platforms such as YouTube, Netflix, Amazon and apps such as Instagram, Musical.ly and many others – including, by the way, a significant quantity of pornographic videos. Let us now consider whether film streaming can reduce greenhouse gas emissions. A study conducted in the USA concludes that online streaming of a 90-minute film saves about one-third of the CO_2 emissions generated by a viewer who drives to a video library 17 kilometres away to borrow a DVD (this being the average distance that Americans used to travel to their nearest video library). The study also finds that the CO_2 emissions associated with renting a DVD through a postal borrowing service are virtually identical to emissions from streaming.[34] Other studies show that the poor performance of streaming can be improved if devices meet the latest streaming standards and server parks make greater use of electricity from renewable sources, which one hopes that they will in future do.[35]

All forecasts indicate that data streaming – and, in particular, mobile streaming by people on the move – will continue to increase significantly; on a global scale, it could triple or quadruple between 2015 and 2020.[36] This growth could eat up the savings arising from improved efficiency standards for devices. There are many reasons why more and more people are likely to stream films: they may do so because it can be cheaper; because it is more convenient; because there is a wider range on offer; because it enables them to access anything from anywhere at any time; or because DVD rental services are going bust and the analogue infrastructure is being displaced by digital systems. Streaming undoubtedly increases the overall consumption of films and music: television and radio audience figures are falling only slightly, which means that the rapidly growing online segment represents additional consumption.[37] At the same time, the volume of data is also increasing – especially for films, because they are being offered in ever-higher resolution. Just as the size of television screens has increased hugely over the years, the volume of data is also growing by a factor of 20 or more in the course of the transition from traditional formats (approx. 4 GB per film) to HD (approx. 8–15 GB), Blue Ray (approx. 20–25 GB) and 4K movies (more than 100 GB). If 3D films become available for streaming, for example as virtual-reality animations, the data transmitted per minute of streaming could increase 40-fold.[38] What are the environmental implications of this huge increase in data traffic? How many additional data centres will need to be built? What resources will that involve, and how much energy?

The situation can be summed up as follows: a shift from analogue to digital devices and applications has the potential to save energy and resources if sufficient use is made of devices such as e-book readers and mp3 players that are purchased new for them to pay off in environmental terms. However, there is a big "but": because online access, such as the streaming of music and films, is becoming so much quicker and easier and often cheaper too, we are demanding more digital services than ever before. In consequence, digitalization in this area is likely to be at best a zero-sum game in environmental terms.

Distributed energy systems call for digitalization

As we have seen, the shift from physical products to digital services often means that material resources are to some extent replaced by energy resources. As we switch to streaming a film instead of buying a DVD or reading the news online instead of purchasing a newspaper, digitalization is steadily electrifying our economic activities and consumption patterns. Of course, the mass of devices and the digital infrastructure still have a material basis, the environmental implications of which – as we have noted – are substantial. But, the further digitalization advances, the more important electricity becomes as a key resource and as the driver of our economy and society.

To become sustainable, energy consumption in all areas of life – not just the electricity sector, but also in the fields of heating and transport – must in future be based on the use of renewable sources. The targets of the Paris Climate Agreement of 2015 specify that, between 2045 and 2060, global greenhouse gas emissions must be reduced to virtually zero.[39] For most industrialized countries in the global North that have in the past been responsible for considerable accumulation of greenhouse gases in the atmosphere, it follows for reasons of climate justice that greenhouse gas emissions from fossil fuels (coal, oil and gas) must be cut to almost nothing in the next ten to 15 years.[40] A shift to renewable sources of energy – wind, solar, water, geothermal, biomass, etc. – is therefore a key component of any environmental sustainability strategy.

We shall see below that the global energy transition will not be achievable without digitalization.[41] To understand the role of digitalization in this, it is helpful to recall how the traditional energy system works. Let us focus on electricity, which is becoming ever more important in our lives. In general terms, the electricity system has in the past functioned on the principle of adjusting supply to demand. For example, if demand was particularly high in the mornings or evenings, the electricity companies produced more at those times – this was achieved in the short term by using pumped-storage power plants and combined-cycle

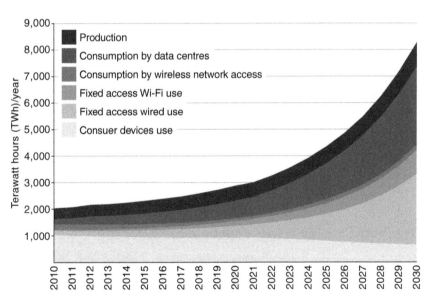

Around ten per cent of global electricity consumption is currently attributable to information and communication technologies. Energy use by end devices is predicted to remain stable or slightly fall in the coming years, but there will be a sharp increase in the electricity used by the cloud (i.e. by data centres) and in the transmission of data.

FIGURE 3.3 Global electricity consumption of ICT

Source: Andrae and Edler (2015).

gas and steam turbines, and in the longer term partly through the use of coal-fired power plants. Characteristic of this system was the use of a small number of large-scale power plants. As the share of renewables in the energy mix increases, more and more electricity is fed into the grid by very large numbers of small, decentralized electricity generation units (wind turbines, photovoltaic systems on house roofs, etc.) rather than by the few large power plants. The logic of the system must thus be reversed, because the wind does not blow nor the sun shine in line with the demand for electricity. In future, therefore, demand must be more strongly aligned with the supply of electricity. "Supply-side flexibility" must be replaced by "demand-side flexibility", and digitalization has a role to play here.[42]

There are two key levers that can be used to manage demand. First, the demand for electricity can be adjusted to the supply. In this context, it makes sense to consider business and domestic use separately. However, there is a limit to the extent to which demand can be manipulated in this way. Second, therefore, there is a need for reserves in the form of storage systems (such as batteries) and power-to-X processes; these are especially important on days that are both calm and overcast, and at night-time. Let us first consider how businesses can "flexibilize" their demand. The idea is simple: if there is surplus electricity, businesses can step up their production or bring forward activities that do not need to be performed at a specific time. For example, cold stores can be chilled to a few degrees below their usual temperature so that chilling does not need to function at times when electricity is in short supply. A somewhat more technically complex approach involves ice storage systems that use surplus electricity to freeze water. When electricity is in short supply, the ice can then be used for chilling without the need for electricity. Heavy industry is also experimenting with pilot projects that make their demand for electricity more flexible. For example, the Trimet aluminium works in Essen, Germany, is planning to introduce a new technology in the form of a "virtual battery" that ensures constant production of aluminium despite varying inputs of electricity.[43] The technique involves an adjustable heat exchanger that maintains a constant energy balance in the smelter despite a fluctuating power supply. This enables Trimet to use more electricity at times when surpluses are available.

Like business users, domestic users can also adjust their electricity use to fluctuations in supply. Where possible, energy-consuming activities – such as operating dishwashers and washing machines, charging laptops and electric cars – can be performed at times when electricity from the sun and wind is plentiful. There is particular potential for this if households generate some of their electricity themselves, for example via solar panels on the roof. Because they are both producers and consumers of electricity, they become "prosumers".[44] And if prosumers have, not only solar panels on the roof, but also a storage battery in the cellar, they can even use the battery to store temporary surpluses of electricity.

If business and domestic users are to adjust their demand for electricity, they must know how much electricity the grid is currently able to supply and how much will be available in the immediate future. Meteorological analysis is already able to provide highly precise forecasts of renewable electricity generation at particular times, and research into new models for predicting the output of weather-dependent sources of energy is ongoing. Information about supply must, however, be accompanied by incentives that encourage electricity customers to use electricity when the supply is plentiful and dial back their usage when less electricity is being generated. Market incentives usually operate via the price, with electricity costing less when sunshine and wind are plentiful than when the weather is cloudy and still. A flexible electricity price that is applied across all relevant customer segments will therefore be a core element of the future electricity system.

But, how will users obtain information about the electricity price and, hence, about the current level of supply? Isn't a flexible electricity system of this sort extremely complicated? Yes, it is – and it is precisely in order to manage this complexity that more digitalization is needed. Both the electricity grid and the electricity market must be controlled digitally; this is where "smart grids" and "smart markets" come in.[45] Significantly more information must be exchanged between different market participants. Millions of business and domestic machines, devices and control units must receive up-to-date information about

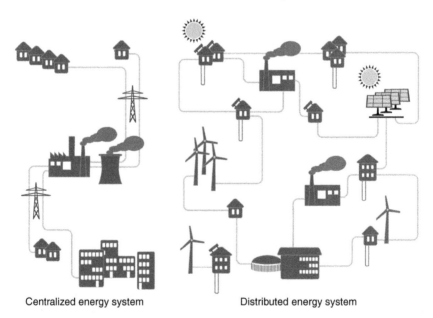

Centralized energy system   Distributed energy system

In a centralized energy supply system, large power plants adjust their supply to the demand from businesses and households. In a distributed system, municipalities and businesses generate their own electricity, sell their surpluses and buy electricity if more is needed. Demand must be flexibilized with the help of ICT to align it with supply at any given time.

FIGURE 3.4   Centralized versus distributed energy system

Source: Diagram based on Egger (2017).

how much electricity is currently available. It is impossible to imagine how this data can be transmitted at frequent intervals to so many participants without digital technology.

The complexity that arises also means that many processes must be automated. A cooling system can turn itself on and off automatically as the electricity price fluctuates; the electrolysis process in the aluminium works or the charging station of an electric car can operate in the same way. Not many people will reduce the cooling power of their refrigerator when the electricity price rises. It is not yet clear to what extent it makes environmental sense to enable domestic devices and appliances to communicate directly with each other and, via a smart meter, with the electricity grid so that the system works. Automating every single charging station for an electric toothbrush in the smart home would be a disproportionate response. For businesses, on the other hand, it will often make sense to automate the manner in which energy-intensive processes are coordinated with the electricity supply.

One aspect of our future electricity system therefore involves increasing the flexibility of demand. But, how can consumption be adjusted to the fluctuations in wind and solar power? What happens when the wind is still for so long that the temperature in the cold store rises too high? And what happens in the opposite case on particularly stormy days when the wind power remains unsold because there is simply too much electricity being generated? This is where we need electricity storage systems and methods of converting surplus power into other forms of energy such as gas and heat – known as "power-to-X". For example, electrolysis can be used to convert electricity into gas (power-to-gas) that can then be used to generate heat or fuel gas-powered vehicles. Power-to-heat involves converting electricity into heat – this is the principle on which heat-pump boilers are based. Battery storage systems and power-to-X are both important elements of the transition to an energy system based entirely on renewables. There is currently much discussion of which of these methods of storing surplus electricity or using it elsewhere is best from an environmental point of view.[46]

Since the energy transition will increasingly involve converting surplus renewable electricity into other forms of energy such as gas, fuel or heat via power-to-X, it is clear that ever-closer links between the electricity, heating and transport sectors will be required – a process that is being termed "sector coupling". When fossil fuels predominated, these sectors were largely independent of each other. Digitalization can help to "couple" the electricity and transport sectors. For example, digital management systems not only enable the batteries of the electric cars of the future or batteries in the cellar to be charged when plenty of electricity is available; they also permit these batteries to function as storage facilities for the grid as a whole that can be drawn on when demand threatens to exceed supply.[47] A management system of this sort must clearly be automated and "smart" communication between the charging stations and the grid is therefore needed. Digital technology can also contribute to the coupling of electricity and heating. For example, a plentiful electricity supply can automatically be used to provide heat for heating systems. Digital automation is particularly appropriate for decentralized systems such as heat pumps.

The use of digital technologies is thus an essential element of the transition to 100 per cent renewable energy. There is no doubt that ensuring that all electricity is "green" is good for the climate. But, what are the environmental implications of manufacturing the devices and infrastructure needed for communication in the smart grid? A study that included a lifecycle assessment of automated home energy management systems – also known as smarthomes – concludes that the energy saved starts to outweigh the energy invested after a period of about 18 months.[48] A lifecycle assessment of smart meters likewise found that the net effect of a smart meter is significantly reduced if the energy needed for production and use is taken into account.[49] Smart management systems and smart meters can certainly have a positive effect on energy consumption, but, in environmental terms, it is likely to be counterproductive to automate large numbers of relatively small household devices so that their use can be geared to the electricity supply.[50]

Furthermore, if houses are fully converted into smarthomes simply in order to maximize the flexible use of electricity, there is a risk that many devices that are still in working order will be replaced and disposed of. It is certainly not necessary to connect every oven and standard lamp to the smart grid. From an environmental point of view, many of the smarthome systems currently coming onto the market need to be viewed with a critical eye. All sorts of systems are available that can control a wide range of aspects of household technology – from activating the alarm system to changing the colour of the living room lighting and turning the television on and off. Only a small number of these inventions are designed to coordinate home electricity demand with electricity generation or reduce demand through smart management. By far the majority of applications are new gadgets that may enhance convenience, but do not necessarily promote the transition to sustainable energy.[51]

A further point to take into account is the fact that automated communication methods and smart digital management systems themselves use electricity that was not previously needed. This is another example of the phenomenon we described at the outset: since digitalization leads to further electrification of our society, it increases the demand for electricity. Indeed, by 2030, digitalization could account for 30 to 50 per cent of global electricity demand (see section "The physical basis of the virtual world"). This results in a dilemma, since, for complete reliance on renewable energies, it is essential that overall demand in countries with high levels of electricity consumption is reduced. For example, the German government's energy strategy assumes that the goal of transition to a 100 per cent renewables-based electricity system by 2050 cannot be achieved unless the current level of primary energy consumption is halved.[52] Yet, the technologies that can make this possible themselves need energy. For the energy system it is therefore a case of settling on a moderate degree of digitalization (with its associated level of electricity consumption), i.e., a "soft digitalization". The principle on which the transition to a sustainable energy system must be based is not "as many smart devices as possible", but rather "as few as are necessary".

Finally, the introduction of smart devices and grids poses two quite different challenges: data privacy on the one hand, and the security and resilience of the electricity system on the other. Flexibilization of demand involves constant communication and recording of which businesses and consumers need electricity when, where and with what devices. The spread of smart devices in the energy system therefore brings with it a host of issues relating to privacy and IT security. The systemic risks of a smart energy system have already been addressed in science-fiction literature – and, in the view of experts, they are more science than fiction. Marc Elsberg's bestseller *Blackout* highlights the risks.[53] In his novel, electricity grids in Italy and Sweden collapse when hackers disable thousands of domestic smart meters. Power cuts ensue: traffic lights no longer work, resulting in numerous traffic accidents. In the days that follow, there are power cuts almost everywhere in Europe – partly because national electricity grids are interconnected. There is no power for almost two weeks. In this seemingly short time, Europe faces its biggest catastrophe since the Second World War. There is no longer any sanitation; epidemics break out. Food supplies run out, water is in short supply and almost all communication systems (internet, telephone, television) fail. And, because public institutions are barely functioning, there are riots and people take the law into their own hands. It is, in short, a horror scenario.

Blackout is, of course, a work of fiction and not a scientific analysis. Yet the two main issues in the book are a talking point even among experts.[54] First, electricity is fundamental to much of our critical infrastructure and its importance is growing – partly as a consequence of digitalization. The electricity grid is becoming *the* critical infrastructure. Second, the digitalization of energy production and the energy market means that the electricity supply can potentially be disrupted via the internet from anywhere in the world. In Elsberg's novel, it is a few cunning hackers who hope that the collapse of Europe and the USA will pave the way for a new post-capitalist civilization. In reality, attacks of this sort can come from all sorts of directions. And they are already occurring: for example, in December 2015 the Ukrainian electricity system was compromised in a cyberattack – thought to be orchestrated by Russia – that disrupted electricity supplies in several outages to 225,000 people across various areas.[55] As a result of this incident and others that have had relatively minor consequences, insurance companies are currently considering the potential risks of cyberattacks.[56] To ensure that the energy system is as resilient as possible – in other words, to protect it against disruption – the number of interfaces between "smart" infrastructure and the open internet needs to be kept to a minimum. This approach has not yet become the norm and, unfortunately, many of the smarthome systems and products currently on the market completely ignore it. Nevertheless, the smaller the quantity of information that is transmitted via mobile phones (apps), local area networks, private routers and so on, the more difficult will malicious intervention become. Wherever possible, the aim should be to set up systems that use encrypted data and operate via wireless networks rather than the internet, so that only authorized users and permitted technologies can access them.

At individual level, the collection, transmission, storage and analysis of the information in smart grids also raise issues in connection with protection of privacy. Applications that are used to manage smarthome devices provide manufacturers with a great deal of data. At first glance, this information may seem unimportant, but it enables holders of the data to draw very specific conclusions about people's daily lives. For example, a study has shown that, on the basis of the high-resolution energy consumption data from smart meters, it is possible to tell when a particular television programme in someone's home is being watched.[57] And smart devices provide much more detailed information than this. In recent years, a number of major IT companies have brought out voice-controlled smarthome devices. Amazon's Echo and similar products from other manufacturers enable users to control electronic devices in the home and do many other things as well, such as play music or make online purchases. The risk with regard to data protection and the private sphere is that Echo records all conversations and transfers them to the cloud. Amazon has recently released this data in connection with the investigation of an offence. In another example, the US intelligence service, the CIA, hacked Samsung smart televisions in order to spy on users.[58]

From an environmental point of view, there is a definite risk that companies will use the vast quantity of personal data to target advertising and product offers with even greater precision, thereby raising the overall level of consumption; this possibility will be examined in more detail in the section below titled "Big data, big needs". A number of other problems arise in connection with privacy law. In extreme cases, smarthome systems can encourage behaviour that has been studied by the sociologist Michel Foucault:[59] because someone knows that he is being constantly listened to and watched in his home, he adapts his behaviour in a sort of pre-emptive obedience and neither does nor says anything that could potentially be used against him. As a smarthome, the home is no longer a comfortable retreat but becomes a veritable "panopticon", a place in which everything is under surveillance.[60] With regard to the technical design of devices and systems and political regulation of them, this means that smart devices and smarthome systems must always be designed in a way that ensures maximum privacy protection. We shall examine this requirement in more detail in Chapter 5.

To enable the transition to an entirely renewables-based energy system, maximize the resilience of the electricity system, minimize privacy risks and, in addition, spread the value created by the new energy system as widely as possible, the aim should be to use appropriately designed digital applications to create an energy system that is as decentralized as possible. In a strongly decentralized energy system, small units – for example, at neighbourhood or urban district level – generate most of the electricity needed by the local community from solar panels and mini-wind turbines, with inputs from larger wind farms or biomass plants nearby as needed. By flexibilizing demand and using storage systems, they can produce much of their electricity independently. A number of studies have already provided detailed descriptions of such systems.[61]

Neighbourhood systems of this sort – also known as micro- or mini-grids – have numerous advantages. With a decentralized structure, it is virtually impossible for the failure of a subsystem to lead to power cuts in other places, since systems are largely independent of each other. In addition, as a result of distributed electricity generation, the electrical installations of the future will belong to many different individuals, municipal bodies, citizens' energy cooperatives and so on. This means that the profits from electricity generation will no longer go to a few large energy companies as at present, but to a wide range of different people (see also Chapter 4). And, finally, decentralized grids provide better protection of the private sphere. Details of individuals' usage habits would not need to go beyond the confines of the small distributed units. This would result in an energy system that truly deserves to be labelled "smart": it is 100 per cent renewables-based, it is relatively resilient to external shocks and attacks, it holds sensitive personal information securely and ensures that users benefit from the proceeds.

A quantum leap for sustainable consumption

Having examined the material basis of digitalization and discussed how it can contribute to the transition to sustainable energy and what pitfalls might be encountered, let us turn to another important area: consumption. Online shopping is now an everyday activity. But, how does digitalization affect which products and services we buy? And how does it influence the overall level of consumption? A social-ecological transformation of our consumption habits must build on changes of two sorts. First, the products and services we consume must be more sustainable. Second, overall consumption levels among the consumers, particularly in the countries of the global North and the rich consumers in the global South, must fall.

For people who want their consumption to be more sustainable, digitalization opens up splendid new options and opportunities. Organic foods are now available everywhere – not just in trendy districts of larger cities, where organic shops have become on almost every street corner. Even in cities, traditional retail outlets selling sustainable textiles may be few and far between, but sustainable textiles can be purchased online without difficulty. Some online marketplaces, such as Made Trade in the USA, Fairmondo in Germany and UK or We Dash Love in Australia, focus entirely on selling sustainable products, providing a genuine alternative to websites such as Zalando. And the fact that we can search for information on the internet makes it easier to compare the conditions under which products of all types are manufactured. With apps such as GoodGuide and Ethical Barcode, consumers can scan a barcode for immediate information on environmental impacts, such as the product's carbon footprint or greenhouse gas emissions, or details of health impacts such as toxicity. For people who want to be caring and sustainable consumers, digitalization represents a real quantum leap.[62]

Furthermore, the internet is the world's biggest flea market: online marketplaces such as eBay and Rebuy enable people to buy second-hand goods and sell little-used products themselves. Swap is the biggest online store for like-new and pre-owned clothing in the USA, Kleiderkreisel is a German platform where people can sell unwanted clothing or buy items from others and, via Freecycle, virtually anything can be swapped or given away. The sharing economy or "peer-to-peer sharing", involving websites and apps that enable people to share things with strangers, also provides a host of opportunities for borrowing or renting goods rather than buying them new. Drivy and SnappCar are carsharing platforms that include comprehensive insurance cover for vehicles that are rented through the site. All sorts of everyday items from beer benches to wetsuits to tools can be shared between private individuals via various local sharing platforms. Peer-to-peer sharing thus has enormous potential for reducing individual consumption and hence saving resources. In addition, a lot of people participate in these schemes because they value the opportunities for social interaction and community involvement that they offer. Many simply find it fun, and it is also a good way of getting to know your neighbours.

And, finally, digitalization paves the way for a possible breakthrough in prosuming: people no longer need to just passively consume what industry manufactures; instead, they can become producers themselves and make their products and services available to others free of charge or on a small commercial scale.[63] In the digital era, the generation of informational content is no longer the preserve of professional journalists, editors, publishing houses and production companies, but, instead, is something that anyone and everyone can be involved in through posts, blogs, tweets or comments; thanks to digitalization, the same change is possible in the world of goods. Home owners with solar panels on the roof can feed their electricity into the grid – or, in the future, when they are linked to micro-grids, supply it to their neighbours. Digital swap shops will make it easier for gardeners and allotment holders to trade their home-grown tomatoes or potatoes with people they don't know. Hobbyists offer construction kits; people who like sewing can sell home-made clothes via Etsy. And open-source companies provide instructions that enable consumers to make their own versions of small items of furniture and other products. In Chapter 4, we explore in greater depth the potential of prosuming for reorienting the entire economy, making it more decentralized and democratic. From the point of view of sustainable consumption, it is already clear that digitalization enables everyone to develop and market their talents without the need for long-distance transport, expensive marketing channels or the advertising industry.

So far, digitalization has offered one thing above all: numerous options. It provides opportunities to avoid buying new products, reduce one's own consumption and opt more often for environmentally sound and fairly produced goods from small-scale commercial production in one's own locality or neighbourhood. But, the general consumption trend shows that society as a whole is not yet making enough use of these options. The examples of good practice still remain in small niches. As a proportion of the general population, the number of people using green apps and platforms and sustainability-oriented websites of all sorts is very small.[64] The sales figures of Made Trade, Fairmondo or Swap are minuscule by comparison to those of Amazon. Prosuming, too, is still a niche phenomenon. And even eBay – the prime example of the internet trade in second-hand goods – is increasingly becoming a marketplace for new items; in 2008, half the goods sold on eBay were second-hand, but, by 2016, the proportion had fallen to 20 per cent.[65]

So, digital sustainable consumption is still a niche phenomenon. But, how is digitalization affecting the overall level of consumption in society at large? Online shopping is clearly increasing, but the usage varies by region. About 80 per cent of internet users in the USA make at least one purchase online, but, in total, only 9 per cent of all retail sales were purchased online in the USA in 2017. In Japan, the share was only 6.7 per cent. In contrast, already in 2016, an estimated 19 per cent of all retail sales in China occurred via the internet.[66] Worldwide, in both percentage and absolute terms, online trade is increasing rapidly year on year. In 2019, retail e-commerce sales worldwide amounted 3.53 trillion US dollars, and these e-retail sales are projected to grow to 6.54 trillion US dollars in 2022.[67] From the point of view of sustainability, this would not be an issue if purchases from over-the-counter retailers were shrinking by a corresponding amount. However, this is not the case. Despite store closing announcements as retailers are dealing with changing consumer habits, the overall industry is in good health. Traditional retailing also continues to grow, albeit slower than the online market. In the USA, retail sales grew between 3.8 and 4.4 per cent in 2018, including a 10 to 12 per cent boost for online and other non-store sales.[68] Despite all the opportunities for prosuming, sharing and re-selling used goods, digitalization is increasing overall consumption. Not enough consumers voluntarily use digital tools that promote sustainable consumption. Many cannot resist the marketing might of the shopping platforms and the seductive power of the smart algorithms; the consumption habits of the majority of the population are too deeply engrained and not readily changed by the availability of green apps and alternative platforms.

And digitalization stimulates, not only online consumption, but also traditional forms of purchasing. Marketing experts describe how online and offline shopping blend to form a "hybrid consumption space" that expands consumption options even further. Smartphones entice customers, not only to shop online, but also to stroll around the shops; they can even encourage consumers to buy while they are standing in the shop. That is bad for the environment in two ways: not

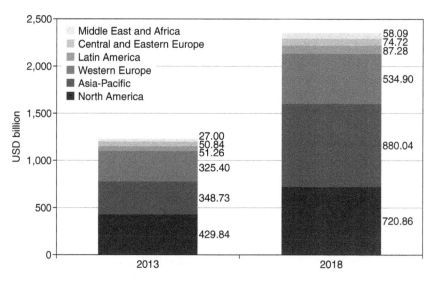

Online shopping is growing rapidly the world over. During the five years between 2013 and 2018, online sales have increased about 70% in North America, Europe, and Latin America and have more than doubled in Asia-Pacific and the Middle East.

FIGURE 3.5   Growth of e-commerce

Source: OECD (2017).

only is overall consumption increased, but setting up and maintaining the duplicate infrastructure – virtual and physical – eats up additional resources. If taking up a sharing offer is just one among a countless number of other consumption options, its transformative potential will vanish.[69] An even greater risk arises from the fact that for a substantial number of people, sharing simply serves to increase the range of consumption options available and enhance individuals' flexibility, but does nothing to reduce the environmentally unsustainable level of consumption. This risk exists, for example, in connection with commercial fully flexible free-floating carsharing, which enables cars to be used simply and on the spur of the moment for single journeys; we shall look at this in more detail in the section "Can delivery services reduce traffic?". Criticism is also being directed at other aspects of the sharing economy, such as platforms that allow private accommodation to be rented for short periods. In many cities, apartment-sharing is being accused of exacerbating an already strained housing situation. And products and services of the sharing economy, of course, leave their own carbon footprint. Although sharing may mean that this footprint is often smaller than that of a hotel room, here again increased consumption can wipe out the savings potential.[70]

Finally, the digitalization of consumption reflects the current economic power structure and existing consumer interests: in niche areas, the opportunities to buy more sustainable products or avoid buying new things are improving, but, in general terms, digitalization is acting as a catalyst of consumption and giving mass consumption a powerful boost.

Big data, big needs

Just how does digitalization stimulate consumption? To understand this, we must look at how users' data is collected and analysed. Most operators of platforms and search engines save and then analyse information about the purchase and use of all types of products and services on the internet – just as they collect information about the streaming of films and music. As an example, let us again consider the purchase of a music album: it is not only the completed purchase process that is recorded, but also all the buyer's surfing activity – perhaps he does a quick search on Bing to find out when his favourite band is going to release its next album, checks the price of the album on otto.de, watches the music video on YouTube and tells friends via Instagram or Snapchat that he is looking forward to getting the album. We leave detailed traces of all our activities on the web. The vast quantity of data thus created is called big data. But this term conceals the fact that the big IT companies create a profile of each user's preferences. This profiling involves collecting every conceivable type of information about us – the work we do, where we live, our purchasing power, health, personal preferences, interests, transport habits, shopping behaviour and also our friends and family and what they prefer and do, etc. This delivers frighteningly accurate statements about our personality. The tracking and analysis of online activity makes everyone a "transparent person". This is used, not only by businesses for personalized marketing, but also by governments and intelligence agencies that have access to this data.

Until now, this issue has been discussed by the media, civil society and politicians mainly from the point of view of privacy. But, profiling does not only intrude on the private sphere and harness confidential personal information – which in itself is worrying enough and can, as we have seen, lead to the manipulation of public discourses or even national elections. Moreover, companies are exploiting people's fundamental right to freedom in order to maximize their profits.[71] Our answers to the question "Who am I?" are increasingly influenced by the fact that companies use our data commercially. For example, Facebook influences our political opinions because it primarily offers us feeds that the Facebook algorithm has identified as fitting our profile. In the same way, shopping platforms offer us "identity-forming consumption" in order to boost their profits. We thus become, not only transparent people, but also malleable people – and compliant consumers. And, as a result of digitalization and big data, the sales-oriented economy is flourishing as never before.

This issue is often glossed over with the comment that businesses simply set out to meet the needs of consumers. But, long before the dawn of the digital age, people were aware that the majority of the consumption "needs" of the "transnational consumer class" – comprising the consumers of the global North and the affluent citizens of the global South – are neither fundamental in nature (food, housing) nor an expression of their deep-seated personal wishes. Instead, they are largely the result of deliberate manipulation by advertising and marketing honchos.[72] From an environmental point of view, this is a serious issue. It is well known that the consumption level of this consumer class has for years been far too high.[73] Analysis of big data and personalization of advertisements create new ways of influencing our buying habits – not just in order to be better able to meet consumers' wishes, but, above all, in order to constantly elicit new wants. Hence, violating privacy not only undermines freedom of speech and democratic institutions, but also contributes to overconsumption and global environmental crises.

Let's take another look at the online streaming of films and music. It enables companies such as Spotify, Deezer, Apple (iTunes), Google (YouTube), Amazon and Netflix in the western hemisphere and Baidu, Tencent and Alibaba in China to offer their users personalized suggestions based on analysis of their previous watching and listening habits. For example, Spotify's "Discovery Playlist" provides each of its 75 million users with a weekly selection of new songs that they have not previously heard, but that match their personal taste in music. While this is without doubt a brilliant way of getting to know new music and films, it is clear that this type of digitalization, which makes media consumption more

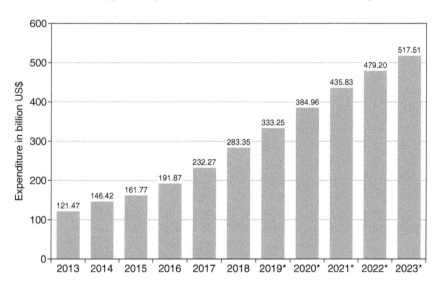

Expenditure on online advertising is rising sharply year on year. In 2017, global expenditure on online advertising has climbed over USD 200 billion. On the revenue side, almost half of this sum was shared between the two companies Alphabet and Facebook in that year. The values for the years marked * are projections.

FIGURE 3.6 Global expenditure on online advertising

Source: eMarketer (2019b).

convenient and puts information at people's fingertips, raises the overall level of consumption and certainly does nothing to reduce the environmental burden. For someone who has been charmed by the "magic" of these algorithms, the question of whether it is better to download mp3 songs rather than buy a CD is simply outdated. Both options seem too rigid to be satisfactory. A constant matching process that takes place in real time and compares highly personal wishes with new and carefully customized consumption offers is far more attractive than the old ways of doing things – but it removes the potential for reducing resource consumption.

When it comes to the consumption of goods and services, companies are less keen to talk about the "magic of the algorithms" that is trumpeted and explicitly sold as a service in the marketing of music and films. This is a smart move, because in relation to goods and services they ought to be talking, not of the algorithms' magic, but of their might and the fact that they use ever more cunning techniques to encourage people to consume. This is most evident in the case of online shopping. It is obvious how online shopping makes information-gathering and purchasing easier and more convenient. Consumers are no longer dependent on opening times; without even moving from the couch, they can find an inexhaustible variety of goods on the internet and can readily compare product features and prices.[74] This service is apparently free of charge. But, we pay with our data, the new "currency" of the 21st century. We are contributing to the big data that businesses use to perfect their marketing. They then press more and more things upon us with ever greater ease – things that we perhaps never even knew about, but that now fill us with longing. Big data in effect gives rise to "big needs".

The analysis of location data is expected to be even more lucrative for marketing purposes than the analysis of search profiles.[75] Precise movement profiles can already be created from the location of a user's mobile phone. In addition, apps such as Google maps (as well as many others that have nothing to do with map or location services) constantly send location data to the provider, thereby enabling companies to deliver advertising that is not only personalized, but also relevant to the consumer's location. For example, a person who is using his or her smartphone to navigate to Starbucks may suddenly get a push message offering a voucher from the competing vendor Balzac. Or, someone who has just been looking for clothes on the internet may get messages about bargains at Zara or Primark just as they approach those particular stores. This stimulates consumption, because the offers are geared, not only to people's personal interests, but also to the situational opportunities for shopping that present themselves.[76]

The personalization of advertising and products is not the whole story. There are many signs that online shopping prices are also being personalized, so that different users are offered different prices.[77] This is already known to occur in the travel industry and in other sectors; there are some studies showing that personalised pricing is already occurring at least to some extent. A recent survey by Deloitte involving over 500 companies found that, among all retailers that have adopted artificial intelligence to personalise consumer experience, 40 per cent of them used AI with the specific purpose of tailoring pricing and promotions in real time.[78] Another study used data of the accounts and cookies of over 300 real-world users in order to test for the presence of price discrimination in 16 popular

Technological utopia: augmented and virtual reality

IT experts predict that augmented and virtual reality could trigger the next major digital revolution.[84] In future, we would wear glasses, not only to compensate for poor vision, but also to add to the reality we see. Information would be projected onto the glasses – this might involve signposting arrows on a navigation app, information about product quality in the supermarket or anything that happens to be useful at that particular moment in our everyday lives. Augmented reality enables us to use apps all the time while keeping our hands free, because we no longer need to be looking at the smartphone display. Augmented reality applications of this sort are already possible, for example on car windscreens. By contrast, virtual reality glasses cannot be used in the street – they transport us into an artificial 3D world in our own home. Glasses from various providers are already on the market. The hallmark of this technology is the perceived "genuineness" of the virtual reality – and this is increasing with every new model.

The potential applications of augmented and virtual reality are wide-ranging and could also change how companies do things. A problematic aspect is the large volume of data that three-dimensional virtual reality involves. Mass use of virtual reality would drastically increase data volumes and require expansion of the IT infrastructure.[85] Although virtual reality films about foreign countries could, in some cases, replace physical journeys, it is far more likely that virtual reality will be used for advertising purposes and thus serve to increase tourism. Both virtual and augmented reality applications will supply the provider companies with highly personal information about aspects of user behaviour such as movement, feelings and finger gestures. If not regulated, this could lead to comprehensive surveillance.[86] Moreover, the information can be analysed and used to improve the personalization of advertising and product offers. With virtual reality apps, we can try on clothes virtually or superimpose furniture into rooms in our home – and then be offered personalized purchasing options. Virtual shopping centres and virtual reality consumption animations in traditional shops – which might highlight special offers or manipulate our purchasing decisions through suggestion – would be likely to increase existing consumption levels even further.

e-commerce websites, of which nine were found to have some element of personalisation. Even though existing evidence of personalized pricing strategies is relatively limited, there are numerous reports from users who have affirmed their encounters with price discrimination.[79]

Personalized pricing again serves to stimulate consumption – while also boosting companies' profits. It is estimated that Amazon changes the prices of around three million products every day, creating an opaque jungle of special offers on the one hand and periodic high prices on the other, voucher campaigns here and discounts there, that even price comparison engines can only partially penetrate.[80] The personal profiles of individual customers are, of course, at the heart of this dynamic pricing: a customer who has often sought out bargains in the past is more likely to be offered a personalized discount than someone who has rarely clicked on cut-price offers. People shopping on mobile devices are often charged higher prices than those who use a desktop computer, while Mac users are being steered to pricier options on a website than Windows visitors.[81] And observations indicate that electronic products tend to be cheaper on Wednesdays than on other days, while shoes are cheaper on Thursdays and beauty products on Fridays. Online shopping in general is more expensive at the weekend than on weekdays.[82] Why on these particular days and not on others? Because analysis of big data shows that this maximizes sales. But, that only applies for the time being: prices are constantly adjusted according to the information that we users generate on the web. The smart algorithms that supply this information are the lubricant that keeps the consumer society going. And the browsers, apps and online platforms that we use are the most cunning salespeople around.[83]

In environmental terms, an even greater risk arises from the fact that the internet companies and the entire advertising industry not only know what we looked at, liked, commented on, rejected, bought or even sold on the internet yesterday. To an ever-increasing extent, they also know what we will be attracted to tomorrow – not just in theory, but in practice. For example, internet advertising doesn't always offer us the same suggestions for products we might like to purchase, nor are alternative related items simply proffered at random. Instead "suggestive algorithms" based on the analysis of big data, personality profiling, trend analysis and so on are able to predict with ever greater precision what we are likely to buy in the future.

Commercial surveillance and monitoring of consumers thus yields all sorts of ways of pushing the already high and unsustainable level of consumption higher still. It is true that "analogue marketing" has already achieved this, but, by comparison with the opportunities presented by digital marketing, the analogue version appears relatively ineffective – indeed, almost innocuous. Advertising columns, billboards, newspaper advertisements and television commercials offer products that many viewers are never going to be interested in and that they therefore simply ignore. In the digital era, by contrast, companies know from big data exactly who we are, where we are right now and what we are likely to be interested in. They can therefore make us the perfect offer at a personalized price. Although this development may not yet have advanced quite as far as some supporters like to claim, personalization is likely to mean that the internet will soon only woo us with products that are quite closely matched to our interests – or that will highly effectively generate future desires that we currently cannot even imagine. This makes it far harder to resist the offer. It seems that the internet, which was originally intended to be an information exchange, has morphed into a giant selling machine.[87]

E-commerce: anything, anywhere, anytime

No matter whether we are in a fast-food restaurant, in the doctor's waiting room, at the bus stop or at home on the settee in the evening: with the smartphone, tablet and PC we can access the largest shopping centre in the world from anywhere at any time. However obscure the desired item, there is no longer any need to pound the pavements: consumers' wishes can be satisfied almost effortlessly in an instant. Automatic payment systems such as PayPal have further increased the ease of online shopping. With its one-click ordering system, Amazon – which is by far the largest online marketplace – has even removed the steps that purchasers perceive as time-consuming. It is now no longer necessary to log in, enter a delivery address and payment method, accept the terms and conditions and so on: simply tap the touchpad and the parcel courier will deliver the desired jeans, games console or underwater camera to your door the next day. And people for whom even tapping is too much can press a Dash Button to re-order a predefined product, or call out their wish to a virtual assistant such as Apple's Siri or Amazon's Echo or Alexa. As well as generating needs through personalized advertising and prices, digitalization also boosts consumption by enabling us to shop round the clock every day of the year, wherever we happen to be.

Digital connectivity not only makes shopping more efficient and convenient, but also opens up entirely new opportunities. For example, the Pinterest app completely reverses the usual sequence of a purchase. Instead of first having to look for a product in a shop or on the internet, users take a photo of an object with their smartphone and upload it, whereupon product recognition software provides details of online shops where they can buy this item or something similar straight away. Pinterest's aim is to make as many items as possible available to consumers in seconds. Pages could be filled with more examples of this sort, but the conclusion is always the same: digitalization can drastically reduce the transaction costs of consumption. Buying products and services is becoming steadily quicker and easier, and often cheaper too. And, generally speaking, the more time, money and effort is saved, the more people consume.

Empirical studies demonstrate that there is a clear correlation between increases in the range of shopping opportunities and the efficiency of the shopping process on the one hand and the level of consumption on the other.[88] For example, a representative survey of the shopping behaviour of 16,000 interviewees in the USA found that "mobile shoppers" who use a smartphone place larger and more frequent orders than online shoppers who use a home computer.[89]

Linking social media with online shopping also increases consumption. The phenomenon is familiar even on traditional shopping trips: when people go to the shops with a group of friends and celebrate shopping as a social event, they buy more. Facebook, Instagram, Pinterest, Twitter and other social networks stimulate consumption in the same way – the only difference being that the spending triggers are ubiquitous and ever-present and do away with the effort involved in making "real" arrangements to meet. The online shops are themselves active in social networks, pushing advertising, special offers or vouchers – all personalized, of course. Because consumers are using the platforms to exchange messages, filter information, recommend products to each other, rate products and so on, sellers and advertising companies know exactly what they need to offer to whom in order to most effectively entice people to shop.[90] A comparative study of consumers in China, Hong Kong, Taiwan, Germany and Italy has demonstrated the influence of social media on clothing consumption and the growth of fast fashion: in Germany, a quarter of all women are now enticed into buying clothes by social media or fashion blogs, while in China an astonishing 72 per cent stated that this was the case. Facebook and Instagram also influence the frequency and duration of shopping and the amount spent: people who follow fashion via social media shop more frequently and spend more time and money.[91]

If rising consumption is undesirable from an environmental point of view, does it at least help people lead happier lives? The bitter irony of the (digital) acceleration of consumption is that it does not even lead to greater satisfaction. The opposite appears to be true: the shorter the purchasing process, the more quickly the satisfaction fades. Of the 40 per cent of respondents who were classed as "excessive consumers" because they buy clothes at least once a week, more than half stated that the satisfaction wears off after a few hours or, at most, within a day. A third actually admitted to feeling even emptier and more unfulfilled after shopping than they did before.[92] Without a detox programme to tackle their shopping addiction and escape the vicious cycle, people will try to tackle the resulting emptiness through even more frequent shopping.

Can delivery services reduce traffic?

The growth of online shopping is viewed with concern by traditional retail outlets. Some even fear that e-commerce could at some point completely replace conventional shopping, resulting in empty city centres.[93] However, in the light of the general rise in the level of consumption that has already been described, these fears seem exaggerated. In prime locations – especially city centres – people are still likely to continue to enjoy going to the shops. Those places will retain their shopping streets, although their appearance will change, with fewer small shops and more showrooms. The risks are greater for medium-sized towns and greatest of all for small communities and rural areas.[94] Food shops selling every-day essentials will have least reason to fear the competition of online shopping.[95] Nevertheless, all towns and municipalities are likely to suffer from the loss of commercial taxes when outlets such as pharmacies, newsagents and shops selling household or electrical goods close down.

While shops slightly decline as e-commerce grows, there is an increase in the number of delivery vehicles bringing our online purchases to our homes. What are the environmental implications of this? Indeed, it can be a good idea – especially in rural areas – to replace countless individual car journeys to the shops in the nearest main town with a combined delivery service. Many studies have looked at the question of whether online shopping can reduce energy consumption and greenhouse gas emissions and thus be more environmentally friendly than personal shopping by car.[96] As when comparing books with e-readers or DVDs with streaming, the answer depends on the specific circumstances – in particular, on whether consumers drive to the shops and, if so, how far they drive, and on shopping habits and settlement structures. Various studies based on different contexts and assumptions conclude that the volume of shopping-related traffic could be reduced by 25–75 per cent if all private shopping by car was replaced by delivery services.[97]

However, shopping is often combined with journeys made partly for other purposes (travelling to work, leisure activities). These journeys cannot be entirely assigned to the carbon footprint of shopping. Besides, it is a widespread habit to purchase online piece by piece. Yet, if several products that were previously purchased together on one shopping trip are delivered in separate parcels, the transport-related footprint of online shopping may be worse than that of the outing to the shops.[98] Capacity utilization of the delivery vehicles is also crucial: one study shows that, for the carbon emissions balance of delivery services to be better than that of individual shopping trips, one truck must deliver to at least 20–40 customers.[99] Conventional shopping in areas of dense settlement structures such as city centres still outperforms online shopping, especially if consumers can shop on foot.[100]

Nevertheless, most studies conclude that online shopping can reduce the traffic volumes and greenhouse gas emissions associated with conventional shopping. But, what happens when consumption levels rise higher still – partly because delivery services make online shopping ever more convenient? It has not yet been studied how increased consumption as a result of digitalization may affect the prospect of traffic reduction.

To expand online shopping, many internet companies are considering how they can remove one of its major disadvantages by comparison with traditional shopping, namely the logistically necessary interval between order and delivery. Shopping from a smartphone or computer lacks the "instant gratification" that results from having the goods in your hands immediately. Fast delivery times and low shipping costs are therefore a key focus of development in digital shopping. Futuristic prophecies frequently proclaim that delivery times are about to be drastically shortened. For example, in 2014 the internet platform eBay announced that, in 2015, its eBay Now scheme – using the slogan "from your phone to your door in about an hour" – would offer same-day delivery in 25 cities worldwide. Yet this scheme had been shut down before it succeeded. Amazon's Prime programme offers same-day express delivery and Amazon is now testing Prime Air, which uses drones to deliver goods from the warehouse to the buyer's home in just 30 minutes. This makes the Estonian company Starship seem almost old-fashioned already: Starship is developing small self-driving robotic delivery vehicles that it claims will soon populate the pavements and replace parcel services. However, none of these "technotopian" ideas could be scaled up yet – there are not only economic, but also legal, issues that must be resolved first, such as the question of liability in the event of damage by autonomous vehicles or drones. From an environmental point of view, these obstacles are probably a blessing, because maximizing the speed of delivery is incompatible with maximizing sustainability: when speed is the priority, products are less likely to be combined into a single order and it becomes more difficult to deliver to several customers at the same time. In addition, ever-faster deliveries do nothing to further the aim of reducing excessive consumption to a sustainable level.

E-commerce with delivery services has a further environmental snag: many purchases are returned. Because clothes cannot be tried on before purchase, items of clothing are often ordered in two different sizes, one of which (or both, if the buyer doesn't like them) is bound to be sent back. About one in ten of all items ordered are returned to the seller; in the case of clothing the figure is one in two. When considering whether online shopping plus a delivery service saves energy and carbon emissions by comparison with personal over-the-counter shopping, the very high rate of returns must therefore be taken into account.[101] Worst of all, Amazon and other online platforms destroy millions of returns every year just because this is more cost-efficient than bringing the still new, and unused, goods back into the market.[102]

As well as generating transport-related emissions, online shopping also devours energy and resources in connection with the packaging of goods. Products often come double-wrapped in plastic film. Fragile goods are frequently buried in polystyrene chips in an oversize cardboard box. And the packaging waste mounts up. The total number of containers and packaging waste generated in the USA grew constantly from 27.3 million tonnes in 1960 to about 77.9 million tonnes in 2015.[103] The amount of packaging waste generated in the EU between 2007 and 2016 was estimated, on average, at 79 million tonnes per year.[104] E-commerce is one of the main drivers of this development. It is true that many of the materials used in the mail-order business are recyclable: cardboard is a case in point, and some plastics and other materials can be recycled too. But, recycling itself uses energy – and the process cannot be repeated ad infinitum.

More sustainable logistics systems such as the reusable packaging developed by the German mail-order company Memo and the Finnish company RePack have yet to catch on in the mass market.

In summary, though, we can say that delivery services can have a better environmental footprint than private shopping trips if deliveries are combined in the transport system and if they replace a large number of individual journeys by car. However, there is a risk that this efficiency potential will be negated by faster deliveries that devour energy and resources and by a further rise in the level of consumption. Policy management is therefore needed to enable the environmental potential to be utilized; we shall look at this in more detail in Chapter 6.

Smart mobility – great opportunities, large risks

Online shopping deliveries are not the only area of transport in which digitaliza-tion can trigger positive environmental effects: transport and mobility in general, including passenger transport, can also benefit. The challenge in the transport sector is to initiate a social-ecological mobility transition – a transition that cen-tres primarily on reducing emissions and energy use throughout the sector, but that also makes transport as socially inclusive as possible and that contributes to more viable cities and rural areas. Electric vehicles will be part of this scenario, but we shall not discuss their pros and cons here. The mobility transition will also aim to shift the modal split so that there is less reliance on motorized personal transport and a greater focus on public and shared-used (mass) transport. The authors of the study by the Global e-Sustainability Initiative to which we have already referred calculate that worldwide greenhouse gas emissions in the trans-port sector could be cut by 3.6 gigatons by 2030; this is about one-fifth of total emissions in this sector.[105] The study sees the greatest efficiency potential with regard to goods production and sale in optimizing logistics networks and delivery transport. On the consumer side, emissions could be reduced above all by home-office working arrangements and carsharing, the study suggests.

But, a degree of scepticism is unavoidable – especially in the light of the above discussions. We are already used to downloading e-books instead of driving to the bookshop, streaming films instead of buying or borrowing DVDs, shopping online instead of going to the shops in our car, skyping and videoconferencing at work and even teleworking – we have been doing all these things for some time, so they should have contributed to some reduction in traffic volumes in recent years. Yet neither in Europe or the USA, nor elsewhere in the global South and North, have traffic volumes significantly fallen; instead, in most countries they have continued to rise.[106]

In particular, there has been a significant increase in freight transport. Infor-mation and communication technologies have, without doubt, considerably improved efficiency in logistics and fleet management in the past two or three decades. Nevertheless, with hindsight, we can see that this has not resulted in an absolute reduction in transport emissions. Instead, efficiency increases have paved the way for further increases in transport as a consequence of rebound effects. For example, the concept of just-in-time logistics, which has shifted stockholding from warehouses to trucks on the road, has improved efficiency in terms of time and costs – but at the expense of increasing energy consumption in logistics.

For the transport sector, it is predicted that – among other things – digitally-based services will further optimize logistics networks and freight vehicle routes; that smart traffic management systems will provide information in real time on traffic jams and slow-moving traffic and help to prevent them; and that smart drivetrain technologies will promote "eco-driving".[107] To what extent will these improvements reduce the environmental impacts of transport?

The German Federal Ministry of Transport provides some interesting answers. It forecasts that freight traffic in Germany will increase by 38 per cent between 2010 and 2030, stating that: "Nevertheless, the existing interconnectivity of the modes of transport must be further optimized if we are to manage the above-average growth in the volume of freight traffic in our country predicted in the forecasts for the period to 2030."[108] In other words, digital optimization of the transport system is intended, not to reduce the volume of transport, but to pave the way for its further expansion. Looking back, too, the Ministry concludes that telematics has already "made a major contribution towards enhancing capacity of all transport infrastructure and in almost all modes of transport".[109] And what is true for freight traffic and logistics also applies to use of the private car. Such explicit statement on the relationship between the digitalization of (road) transportation and the kilometres driven and fuel consumed can, most likely, be transferred to other countries as well: in the transport sector, it is highly likely that digitalization is not, at present, contributing to a social-ecological transformation; instead, every optimization of the status quo is furthering the continuation of unsustainable growth trends.

The most radical changes in the transport sector as a result of digitalization are expected in the field of self-driving vehicles. Whether automated vehicles will become widespread and, if so, how quickly, is still unclear – especially as the technology is not yet sufficiently far advanced to operate safely in complex urban traffic. Moreover, there are still a number of legal issues to be resolved before they are permitted on the roads.[110] Yet, if the hurdles are overcome, the technology could become a fixed component of the transport system – with serious consequences for people and the planet. Let us first consider the private car.

Because driverless cars remove the need for people to think about driving, they could become another living space alongside the home and the workplace. Driving is then no longer just a means of getting from A to B as quickly as possible, but, instead, can provide a productive space. Between two appointments, one could be driven around the city for a while and use the time to prepare for the next meeting, hold a video conference, shop online, watch a film, play computer games or simply rest. Longer journeys to work are likely to become even more popular than such round trips that involve using one's car as a living room on wheels. If the car becomes an extension of the office, spending an hour commuting no longer seems such a waste of time as it does now. Alongside these risks of increased traffic, there is a risk that automated driving will result in private transport competing increasingly with more environmentally sound public transport. Many people already find it more convenient to travel by private car than by bus, tram or underground. If they can do this in future without having to drive themselves, they will be able to work or consume while travelling. Automated motorized private transport could then marginalize other forms of transport. This would not be an intelligent use of resources – at least in cities.

The introduction of autonomous cars is currently being pushed by (at least) two financially strong sectors. First, car manufacturers are trying to use digitalization to safeguard the future of the motor car in cities in the 21st century. They want to maintain or even increase the number of cars sold. For them, digitalization comes at just the right time, because the motor industry is facing an existential crisis: caps on air pollution, climate change policies that increase costs and the latest series of fraud scandals threaten to topple the motor car from the central place that it has occupied in our society. In addition, younger people are changing their priorities: they value flexible and multi-modal transport options and want to minimize hassle (such as that involved in finding a parking space) and are therefore starting to abandon the vision, so widespread among their elders, of owning their own car.[111] In that context, the prospect of the new luxury of being chauffeured by robots is an attractive counter-narrative that could re-ignite the public's love affair with the motor car, with all its harmful environmental and social impacts.[112]

Second, autonomous cars are also being pushed by ICT companies and major digital platform providers. They want to provide the software, not only to control the cars themselves, but also for their extensive digital fittings. The occupants must not get bored during the journey! Tesla, Uber, Google, Apple and other companies that are investing heavily in self-driving car technologies will also provide the office, entertainment and gaming facilities with which they will be equipped. Added value will in future be created, not only by selling the car body or at the fuelling point, but also during the journey. The aim of the ICT companies is to have travellers use as many services as possible and thus be exposed to the maximum quantity of advertising. And, of course, they want to collect passengers' location profiles and user data in order to analyse them and convert them into cash. Initial estimates predict that the mobility data of private vehicle users could be worth between €350 and €650 per person per year.[113] Multiplied by the number of cars, this reveals a market worth several billion.

The environmental problems created by self-driving cars are not limited to increased transport volumes and increased consumption, potentially triggered by in-car internet use. A further problem is that self-driving cars require the creation and operation of a huge data infrastructure. In the first place, the cars are constantly scanning their surroundings: this involves 20–60 megabytes of data per car

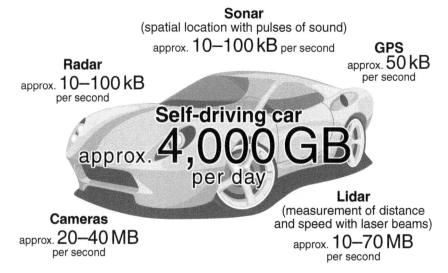

Sonar
(spatial location with pulses of sound)
approx. 10–100 kB per second

GPS
approx. 50 kB
per second

Radar
approx. 10–100 kB
per second

Self-driving car
approx. 4,000 GB
per day

Lidar
(measurement of distance
and speed with laser beams)
approx. 10–70 MB
per second

Cameras
approx. 20–40 MB
per second

Self-driving cars generate an enormous quantity of data, in particular via their cameras and lasers. They also need huge inputs of data for 3D mapping. Rough estimates indicate that the data volume could amount to around 4,000 gigabytes per car per day.

FIGURE 3.7 Data intensity of self-driving cars

Source: Krzanich (2016), Vieweg (2015).

every single second. To this must be added other data flows for GPS, radar and the laser sensors that measure distance and speed. And, lastly, self-driving vehicles not only produce data: they also need a constant input of 3D information about the road environment that they can constantly compare with what they are scanning. According to predictions, the total volume of data could amount to 4,000 gigabytes per vehicle per day.[114] This would mean that the quantity of data generated by just two million self-driving cars would be equivalent to that generated by all of the world's population being online today!

Not only Google, but also other companies such as Uber, Carmera, TomTom and Here (which is backed by Audi, BMW and Mercedes) have already begun to survey the USA and other countries in 3D. Once created, up-to-date 3D maps open up new business opportunities. For example, they can be sold to cities and local authorities for smart-city strategies or as infrastructure for the transport systems of tomorrow. From an environmental point of view, expansion of a vast wireless broadband infrastructure, such as the 5G network, gives cause for concern, as does the enormous computing and transmission capacity of the data centres needed to stream 3D information for millions of vehicles. The environmental footprint of self-driving cars will not be determined only by the energy and resources that go into the manufacture of the car bodies and engines, their digital equipment and, of course, the fuel or the electricity needed for charging. It

Technological utopia: taxi drones

Something of which people have long dreamed could become reality: we could see self-flying "cars" that take us anywhere we want to go and make traffic jams a thing of the past. Drones are already capable of delivering parcels and other products. Established companies such as Boeing and Airbus and start-ups including Ehang, Volocopter and Lilium are now working on flying vehicles that can also transport people. The first air taxi has been undergoing trials in Dubai since the autumn of 2017.[116] In a "brave new world", self-flying drones could collect us from our home in the morning, take us to work, spend a while delivering goods to shops and pick us up after work to take us to visit friends, go to a restaurant or play or watch our favourite sport.

However, uncertainties still surround the practicalities of taxi drones and the feasibility of introducing them on a large scale. It is also unlikely that they would help to make transport more environmentally friendly. Flying usually consumes far more energy than movement on land. It is unclear whether all drones can be powered by renewable energy and what the environmental impacts of the battery technology would be. Furthermore, for the foreseeable future, airborne transport is likely to be open only to the wealthy. A socially inclusive mobility transition will not seek to convey people ever more quickly to more and more distant destinations at which they will spend ever-shorter periods of time. It will, instead, make people mobile by expanding local supply structures and ensuring that, where possible, they can be accessed without any external energy input – such as on foot or by bicycle. Instead of slowing things down sustainably, taxi drones could further speed up our economy and society in resource-intensive ways.

is also necessary to take account of the resources used in setting up and operating the upstream and downstream infrastructure and digital services without which the vehicles cannot function.

Let us draw some preliminary conclusions with regard to the impacts of digitalization on freight transport and motorized private transport. For freight transport, any significant reduction in transport emissions by way of digitalization seems highly unlikely. The digital optimization of delivery transport has already led to rebound effects, partly as a result of cost reductions, and hence to an increase in the kilometres driven. Why should that be any different in future? In connection with private cars, too, the increased convenience that results from digitalization has the potential to push a further increase in traffic volumes. Almost all studies of how self-driving private cars will affect future traffic volumes conclude that the total distance driven will increase significantly.[115] This would have an adverse impact on quality of life, especially in cities. And self-driving cars and digital applications to improve capacity utilization of the roads pose a significant risk to the world's climate.

The environmental risks of private and commercial vehicular transport contrast with the major opportunities presented by digitalization in the field of public and shared-use transport. Indeed, after decades of striving by environmentalists and politicians to encourage greater use of public and shared-use forms of transport, digitalization could trigger the breakthrough they have been waiting for. An example is the widespread expansion of carsharing schemes in towns and cities. Several independent studies have shown that greenhouse gas emissions can be reduced by about 20–25 per cent if station-based carsharing replaces the use of private cars.[117] Apps make it easier to rent shared station-based cars and make schemes more flexible. Digitalization gives rise to additional options, such as peer-to-peer sharing of private vehicles via platforms such as Drivy and SnappCar.

In recent years, digitalization has also led to the creation of free-floating carsharing systems in cities; such systems include Car2Go, DriveNow and SpotCar. An app shows users the location of the nearest free vehicle, which they can then book by mobile phone. It is not clear whether free-floating carsharing has the same potential for savings as station-based systems. For one thing, the more extensive digital infrastructure that is required must be factored into the equation. Much more significant, though, is the fact that free-floating carsharing does far more than station-based carsharing to increase the availability of automobile transport in cities. Free-floating schemes make motorized private transport even cheaper and more attractive and open it up to a broader mass of people – especially for occasional trips for which station-based carsharing is less suitable. In theory, these schemes make it easier for people to do without a car of their own, but research has shown that users of free-floating systems – unlike the members of station-based carsharing schemes – tend to use carsharing in addition to owning their own vehicle, seeing it as a means of increasing their flexibility and maximizing convenience.[118] Moreover, free-floating carsharing competes with mass public transport because it enables people to decide on the spur of the moment to rent a nearby car rather than walk to the nearest bus or tram stop in a way that station-based carsharing does not. There is, therefore, a risk that free-floating carsharing will contribute to an increase rather than a decrease in urban road traffic.

What are the likely impacts of future systems involving self-driving carsharing vehicles or self-driving taxis ("robo-taxis")? For users, there would be very little difference between the two systems: both robo-taxis and self-driving carsharing vehicles would be available on demand. On the one hand, these systems could promote the breakthrough of shared-use (automobile) transport – and perhaps even have the potential to completely replace the currently dominant paradigm of private car ownership. If a vehicle collects us at the door and conveys us autonomously to our destination – with no need for us to look for a parking space and so on – it would win hands-down over the private car for sheer convenience. Optimistic scenarios therefore assume that such systems are likely to be so popular that the number of cars in cities could fall significantly; self-driving privately owned cars are not expected to produce such an outcome. It is estimated that city-wide introduction of robo-taxis could reduce vehicle numbers by 70–90 per cent. On the other hand, the same studies show that traffic volumes could still increase by 20–40 per cent.[119] This is partly because there would be more empty journeys, but also because the interconnectivity of vehicles would improve traffic flows, thereby preventing traffic jams and thus making driving more attractive. There is also a risk that self-driving cars would compete strongly with public transport and travel by bike or on foot.[120] Although the fear that self-driving cars could completely replace public transport is probably exaggerated, the end result of the combination of self-driving carsharing vehicles and self-driving taxis would likely be an increase in energy consumption and transport emissions despite shared-use vehicles. Opportunities to reduce the number of traffic lanes, thereby creating more space in cities, would also be removed. And, finally, the consumption of resources by the ICT infrastructure would increase. A rise in traffic volumes is particularly likely if flat-rate tariffs of the sort used by Spotify and Netflix become the norm. The providers of self-driving carsharing schemes could make money from flat rates if the majority of their profit was no longer generated by the distance driven, but came instead from the marketing of digital services during the journey and the analysis of user data – with the associated problems of protection of the private sphere and the risk of surveillance.

While we are critical of the potential of self-driving private cars, robo-taxis and driverless carsharing, self-driving minibuses could improve local public transport networks in ways that are environmentally sound and make sense in terms of urban planning – provided that they complement bus, tram, underground and rapid transit networks and are not intended to replace them. The opportunities become clear if we consider one of the disadvantages of local public transport at present – the problem of the "first and last mile". People are not conveyed direct from door to door, but must get to and from the bus, tram or train stop on foot, by taxi or by free-floating bikesharing. Self-driving minibuses could close this gap. They would not have any fixed stops, routes or departure times, but would collect passengers and take them direct to the appropriate stop. The key point is that, in contrast to robo-taxis or driverless carsharing, several passengers travel in one vehicle. Even more importantly, the self-driving minibuses are integrated into the mass transport infrastructure. This has the potential to reduce, not only the number of parked cars, but also the number of vehicles on the roads and the distances driven.[121]

Digitalization can contribute to a social-ecological transformation of transport systems in rural areas as well as in towns and cities. Particularly promising are interconnected systems that promote access to shared-use transport facilities. The majority of journeys between villages and small towns, whether for leisure, shopping or work-related purposes, involve only one person per vehicle. In this context, better utilization can be achieved relatively easily: offers of lifts to the nearest shops or to the park-and-ride site can already be made and taken up via apps such as Mobile Together. Such schemes could be heavily promoted by local authorities; there is considerable scope for expansion here. In addition, digitally-based bus-on-demand systems can improve local public transport by making it easier to book telebuses and combine them with other forms of transport, especially if it becomes possible to use self-driving minibuses for this purpose. At present, bus-on-demand services are available in only a few places and they often function only moderately well, because capacity planning is difficult and the system is expensive to operate. Digitalization could improve both these aspects and bring a new springtime for public and shared transportation.

Finally, measures that are designed to change the behaviour of transport users and encourage a shift from motorized private transport to public transport could make a significant contribution to transformation of the transport system. Multi-modal transport apps such as Moovel in the USA, RailYatri or Chalo in India, Citimapper in the UK, Chelaile in China and Transit, which is used in multiple cities worldwide, represent a first step in this direction.[122] They enable users to combine different means of public transport, track public transportation on a map-based view in real-time, get delay updates, traffic information and can be booked on the go. Findings from a survey and the analysis of key characteristics of 83 transportation apps in the USA, suggest that multi-modal app users do change their travel behavior in response to information provided and, hence, contribute to a reduction in vehicle use.[123] To achieve connected intermodal mobility of this sort, the framework for the use of open data platforms that cover all modes of transport must be improved. And the "roaming" system that has become the norm in the EU's telecommunications market should in future also apply to transport. In all European countries, there are many different mobile phone networks – such as Deutsche Telekom, France Télécom, Telekomunikacja Polska and so on – but anyone with a mobile phone contract can now (and often without even noticing it) make phone calls over foreign networks at no extra cost. The message makes its way from the sender to the recipient, no matter which network is used as the "carrier".

In transport, by contrast, travellers going from A to B must decide on a means of transport as the "carrier". A passenger travelling with Deutsche Bahn, British Rail or Trenitalia cannot automatically pick up a rental bike at the end of her rail journey in Berlin, London or Rome; someone travelling by tram cannot simply use a carsharing scheme for the last mile, but must book it separately. "Roaming" that covers all transport operators and service providers could make this easier. Local public transport could blossom if, with a single click, one could, for example, buy an integrated ticket for bikesharing from one's home to the metro station, an underground journey to the outskirts of the city and then carsharing for the last mile – and if, in addition, all the modes of transport and timetables were intelligently coordinated with each other and the changes from one mode of transport to another were displayed in real time on one's smartphone. There would then be no need to buy and maintain one's own car and find somewhere to park it. The costs of the public and shared-use transport would be relatively low and, above all, digitalization provides public transport with the flexibility and convenience that could give it an edge over motorized private transport.

Let us draw a second interim conclusion about the impacts of digitalization on public and shared-use transport. Digitalization provides some promising opportunities for shifting the modal split away from motorized private transport and towards public and shared-use modalities. Furthermore, the digitalization of public transport and the nationwide availability of various sharing schemes (involving bikes, e-scooters, cars, etc.) open up the prospect of transforming the traditional infrastructure and mobility habits of the automobile society and the "car-friendly city" through new transport infrastructure that will make cities fit for the future. The time savings and cost reductions resulting from the switch from private to public modes of transport may lead to rebound effects. Nevertheless, digitalization may prove to be a key tool for improving quality of life, especially in cities, and meeting the challenge of providing people with sustainable means of transport, while at the same time tackling global climate change.

Let us conclude with an overall verdict on mobility. What conclusion can we draw from the opportunities and risks associated with the digitalization of motorized private transport and freight transport on the one hand and of public and shared-use transport on the other? Uncritical approaches, such as that reflected in the slogan "Digitalization first, doubts second" that was used in the latest German parliamentary election campaign, will not contribute to sustainability. Instead, we must think hard and manage things wisely while a range of opportunities are still open to us. Once a digital mobility system has become established, paving the way for a social-ecological transition of the transport system is likely to become significantly harder. Technical savings potential will not be realized unless smart mobility schemes are backed up by smart transport policies that significantly reduce traffic volumes and switch the remaining traffic to more environmentally sustainable modes.

At the heart of a mobility transition should be a public transport system that uses digitalization to convert its previous disadvantages into advantages. As this public transport system is progressively digitalized, motorized freight and personal transport should not be stepped up through digital optimization, but should be made slower and more expensive. In Chapter 6, we shall describe how this can be achieved. With regard to self-driving cars, policy management could involve permitting them only as an add-on to, or in combination with, public local transport and the railways. Self-driving minibuses can be promoted, while restrictions should be placed on robo-taxis and carsharing schemes involving self-driving vehicles; those restrictions can easily be formulated as conditions in licenses for transport providers. At the same time, steps should be taken to encourage the majority of urban journeys to be undertaken by mass transport and not by car or self-driving minibus. This can be achieved via pricing structures or via an algorithm that prohibits public transport routes being replaced by (self-driving) cars.

Industry 4.0: more efficiency, more growth

In recent years, a new vision of industry has been propagated in economies of the global North under the catchphrase industrial internet of things or "Industry 4.0". This next industrial revolution could be the first to be proclaimed before it occurs. It follows that it should be seen less as a scientific concept than as an industrial policy project. For the last few years, the idea has been pushed by numerous industrialists and economic policy-makers the world over.[124] The German Federal Ministry for Economic Affairs defines Industry 4.0 as follows:

> When components communicate with the production equipment by themselves, order a repair to be undertaken when needed or new material to be bought – when people, machines and industrial processes are intelligently networked, then this is what we call Industry 4.0. After the emergence of the steam engine, the conveyer belt and computers, we are now facing the fourth industrial revolution, which is all about smart factories.[125]

Thus, at the heart of this new wave of industrial development is the networking of machines and autonomous communication between them, and also the networking of machines with people and things. This will result in what are known as "cyber-physical systems". Cyber-physical systems include, not only the networking of machines and things via (micro)computers, but also the use of RFID (radio-frequency identification) chips and sensors, analysis of the big data thus obtained, process management via the cloud and the introduction of "additive production processes" such as 3D printing. One could, therefore, say that the industrial internet of things is the widespread digitalization of production.

Many companies strong in mechanical engineering aim to increase their economic growth and competitiveness and, accordingly, countries strong in the producing industries hope to advance their place as a location for business and industry. The potential of growth by implementing the industrial internet of things is predicted to generate $15 trillion of global GDP by 2030.[126] Several additional studies have been developed at the country level. For instance, according to a report by the Boston Consulting Group, the industrial internet of things in Germany will contribute about 1 per cent annually to economic growth (GDP) and drive additional revenue growth of €30 billion per year. The study predicts that the process will generate investment of €250 billion and create 390,000 new jobs.[127] This economic scenario is based on the consultants' expectation that the industrial internet of things in Germany could result in productivity gains of 5 to 8 per cent over ten years.[128]

It, therefore, seems likely that this growth will lead to increased consumption of resources and energy – after all, more will be produced, and the investment in new technology and infrastructure will also need energy and resources. Accordingly, some experts regard a shortage of resources and rising energy prices as the biggest challenges of the industrial internet of things.[129] Sensors, for example, will need a variety of metals that are in short supply, including tin, tungsten, platinum and tantalum (see also the section "The physical basis of the virtual world?"). And the demand for sensors will rise sharply with the growth of the internet of things. Silver, copper and aluminium are needed for the manufacture of RFID chips. In 2010, the number of RFID tags sold worldwide exceeded two billion and, for 2019, the number is expected to rise to 20 billion tags versus 17.5 billion in 2018.[130] In the past, medium- to long-term estimates of the markets for RFID have often been exaggerated, but even "moderately" rising production could lead to over-exploitation of the available resources – with the accompanying environmental burden of mining, transport and the manufacturing process.[131] We have already described how digital technologies could drive up global demand for selected resources over the next 20 years.

Quite apart from the expected boom in the output of the economy and in the demand for resources, supporters of the industrial internet of things believe that digitalization could make the German economy more environmentally sustainable.[137] This view is based on the assumption that the increases in energy and resource efficiency would be bigger than the expected growth. The expectation is that the additional resources needed to manufacture the technologies would be more than offset by the major savings that would arise from large-scale application of these technologies, resulting in a net reduction of environmental impact.[138]

However, even though the industrial internet of things have been under discussion for some years, no studies – let alone scientifically robust ones – have yet been published to confirm this in any way.[139] The optimistic statements of business leaders and politicians are, therefore, no more than hopes – or even greenwashing. The industry-led study by the e-Sustainability Initiative that has already been cited – one of the few that quotes figures at all – calculates that, by comparison with a business-as-usual scenario, the digitalization of industrial processes

Technological utopia: the internet of things

In future, this utopia suggests, everything will be connected to everything else. In our business and personal lives, we will be surrounded by billions of sensors and smart devices.[132] This huge internet of things will encompass, not only industrial machines, but also heating systems, household appliances and entertainment electronics – and even connected T-shirts and "smart" socks.[133] The coffee machine will spring into life when the alarm clock rings. The refrigerator will order supplies from the supermarket to stop you running out of milk. The computer will provide storage space for the calculation of bitcoins. According to the utopian vision, the internet of things will enable many processes to be automated, removing the need for active action and decision-making by users.

In some cases, such applications are already improving energy efficiency. For example, smart energy management systems only heat up hot water when users approach their home with their smartphone. Lights automatically go out when you leave the room. The more renewable the electricity becomes, the more important it will be to automate demand, which needs to be matched to the fluctuating supply from renewable sources (see section "Distributed energy systems call for digitalization"). However, many devices in the internet of things themselves consume electricity. Scientists, therefore, warn that the overall result could be a significant increase in energy consumption.[134] There is no doubt that the internet of things also entails a sharp increase in demand for material resources – with all the social and environmental problems associated with mining. Roughly 15 billion digital devices were in the market in 2015, a third of which are associated with the internet of things; for the year 2022, this number is projected to climb to more than 27 billion devices, more than half of which will be part of the internet of things.[135] In connection with privacy, too, we must ask who will receive all the sensitive information about individuals' detailed usage that the sensors of the billions of devices will send to the cloud.[136]

could save more than 1.2 gigatonnes of CO_2 by 2030.[140] But, as has already been shown in the section "Smart mobility – great opportunities, large risks" the study has some weaknesses. It specifies neither where the data comes from nor what the assumptions are based on. Even more seriously, it takes no account either of the rebound effects or of the resources needed to provide the digital devices and infrastructure. Yet, as many examples in the preceding sections have already shown, these aspects of consumption can be so significant that they offset or even exceed the savings potential of efficiency improvements.

Indeed, rebound effects in manufacturing could be particularly high. Many studies show that energy and resource efficiency increases in industry often result in large rebound effects.[141] Moreover, the industrial internet of things will increase, not only energy and resource efficiency, but also the productivity of labour and capital. This means that, beside the rebound effects, there may also be macroeconomic growth effects in businesses and sectors.[142] As a result, the digitalization of industry has the potential to create a surge in the manufacture of cars, electrical devices and appliances, machines and other products. In economic terms, this may be precisely the desired effect, but, from an environmental perspective, the fact that the increased demand wipes out the savings potential of the resource efficiency improvements and, hence, the reduction in environmental impacts is disastrous. The relationship between resource efficiency and resource demand is such that, for each unit of value – each ball-point pen, printing machine or car that is made – fewer resources and less energy are required, but, because significantly more units are produced, the overall result is likely to be, not a decline in the demand for energy and resources, but a continuously high level of demand, or even an increase.

We shall return to these conclusions about the environmental impacts of digitalization and pursue them further in our discussion of labour productivity and jobs. In Chapter 4, we discuss whether increases in the productivity of labour as a result of automation and robotization can realistically be expected to lead to new jobs or whether they will result in increased unemployment. When work is replaced by digital technology, new jobs can only be created at the same time if high growth in output results in a net growth in jobs.[143] And growth is, without doubt, the primary aim of the ambitious campaign for the industrial internet of things.

Notes

1 Bundesministerium für Wirtschaft und Energie, Industrie 4.0, 2019.
2 See e.g. Matsuda/Kimura, Usage of a digital eco-factory for sustainable manufacturing, 2015; Weber et al., Foresight and technology assessment for the Austrian parliament — Finding new ways of debating the future of industry 4.0, 2018; and more.
3 GeSI/Accenture, Smarter 2030. *ICT Solutions for 21st Century Challenges*, 2015.
4 For slightly different, but more robust, calculations see e.g. Schien et al., Modeling and Assessing Variability in Energy Consumption During the Use Stage of Online Multimedia Services, 2013.
5 For comprehensive introductions to the issue area see e.g. Hilty, *Information technology and sustainability*, 2008; Hilty/Aebischer, *ICT Innovations for Sustainability*, 2015a; Hilty/Aebischer, *ICT for Sustainability*, 2015b; Horner/Shehabi/Azevedo, Known unknowns, 2016.
6 Manhart et al., Resource Efficiency in the ICT Sector, 2016.
7 Chan et al., Dying for an iPhone, 2016b; Chancerel et al., Estimating the quantities of critical metals embedded in ICT and consumer equipment, 2015; Pilgrim et al., The Dark Side of Digitalization, 2017.
8 Balde et al., E-waste statistics, 2015.

9 Greenpeace, From Smart to Senseless, 2017b.
10 The Shift Project, Lean ICT, 2019.
11 Koomey *et al.*, Implications of historical trends in the electrical efficiency of computing, 2011.
12 Roser/Ritchie, Technological Progress, 2017.
13 Coroama/Mattern, Digital Rebound, 2019; Hilty, Why energy efficiency is not sufficient, 2012; Santarius/Walnum/Aall, *Rethinking Climate and Energy Policies*, 2016.
14 Apple, iPhone 3G Environmental Report, 2009; Apple, iPhone 7 Environmental Report, 2017.
15 Andrae/Edler, On Global Electricity Usage of Communication Technology, 2015; Salahuddin/Alam, Information and Communication Technology, electricity consumption and economic growth in OECD countries, 2016; for numbers for Germany, see Fichter *et al.*, Gutachten zum Thema "Green IT-Nachhaltigkeit" für die Enquete-Kommission Internet und digitale Gesellschaft des Deutschen Bundestages, 2012; Stobbe *et al.*, Entwicklung des IKT-bedingten Strombedarfs in Deutschland. Abschlussbericht, 2015.
16 de Decker, Why We Need a Speed Limit for the Internet, 2015.
17 International Energy Agency, *World energy outlook 2016*, 2016; Mills, The Cloud Begins With Coal, 2013.
18 Greenpeace, Clicking Green, 2017a; see also Hintemann/Clausen, Green Cloud?, 2016.
19 On obsolescence see e.g. Oehme *et al.*, *Strategies against obsolescence*, 2017; Proske/Jaeger-Erben, Decreasing obsolescence with modular smartphones?, 2019.
20 Goleman/Norris, How Green Is My iPad?, 2010; Ritch, The Environmental Impact of Amazon's Kindle, 2009.
21 See also Bull/Kozak, Comparative life cycle assessments, 2014; Moberg *et al.*, Printed and tablet e-paper newspaper from an environmental perspective – A screening life cycle assessment, 2010.
22 Chan/Fung/Overeem, The Poisonous Pearl, 2016; Manhart *et al.*, Resource Efficiency in the ICT Sector, 2016.
23 Für einen Überblick see Bull/Kozak, Comparative life cycle assessments, 2014.
24 The Green Press Initiative, Environmental Impacts of E-Books, 2011.
25 Kozak/Keolelan, Printed scholarly books and e-book reading devices, 2003.
26 A conclusion also drawn by Jeswani/Azapagic, Is e-reading environmentally more sustainable than conventional reading?, 2015.
27 E.g. Achachlouei/Moberg, Life Cycle Assessment of a Magazine, Part II, 2015; Ahmadi Achachlouei *et al.*, Life Cycle Assessment of a Magazine, Part I, 2015.
28 See Verband privater Rundfunk und Telemedien, Grafiken zur Mediennutzungsanalyse 2016, 2016.
29 Türk *et al.*, The environmental and social impacts of digital music, 2003.
30 Weber *et al.*, The Energy and Climate Change Implications of Different Music Delivery Methods, 2010.
31 Apple, Environmental Report iPod Touch (6th Generation), 2015.
32 Hogg/Jackson, Digital Media and Dematerialization, 2009; see also Malmodin *et al.*, Greenhouse Gas Emissions and Operational Electricity Use in the ICT and Entertainment & Media Sectors, 2010.
33 Cisco, The Zettabyte Era, 2016.
34 Shehabi *et al.*, The energy and greenhouse-gas implications of internet video streaming in the United States, 2014.
35 Seetharam *et al.*, Shipping to Streaming, 2010; Sivaraman *et al.*, Comparative energy, environmental, and economic analysis of traditional and e-commerce DVD rental networks, 2007; see for a general presentation Stobbe *et al*, Entwicklung des IKT-bedingten Strombedarfs in Deutschland. Abschlussbericht, 2015.

36 Cisco, The Zettabyte Era, 2016; Ericsson AB, Mobility Report: Traffic Exploration, 2017.
37 Horowitz Research, *State of Cable & Digital Media*, 2016.
38 Cisco, The Zettabyte Era, 2016.
39 Rogelj *et al.*, Energy system transformations for limiting end-of-century warming to below 1.5°C, 2015.
40 IPCC, *Global warming of 1.5°C*, 2018; UNFCCC, Paris Agreement, 2015.
41 See the Act on the digitalization of the energy turnaround in Bundesregierung, *Gesetz zur Digitalisierung der Energiewende*, 2016; see also Zimmermann/Wolf, Sechs Thesen zur Digitalisierung der Energiewende, 2016.
42 See e.g. Alstone/Gershenson/Kammen, Decentralized energy systems for clean electricity access, 2015; Vezzoli *et al.*, *Designing Sustainable Energy for All*, 2018; Xu, The electricity market design for decentralized flexibility sources, 2019.
43 Nübold, "Virtuelle Batterie" der TRIMET wird Teil der KlimaExpo.NRW, 2017.
44 European Parliamentary Research Service, Electricity 'Prosumers', 2016; Toffler, *The Third Wave*, 1980.
45 See in detail e.g. Ipakchi/Albuyeh, Grid of the future, 2009.
46 See e.g. Blanco/Faaij, A review at the role of storage in energy systems with a focus on Power to Gas and long-term storage, 2018; Koj/Wulf/Zapp, Environmental impacts of power-to-X systems, 2019; Sternberg/Bardow, Power-to-What?, 2015.
47 See also Kittner/Lill/Kammen, Energy storage deployment and innovation for the clean energy transition, 2017.
48 van Dam/Bakker/Buiter, Do home energy management systems make sense?, 2013.
49 Malmodin/Coroama, Assessing ICT's enabling effect through case study extrapolation, 2016.
50 Louis *et al.*, Environmental Impacts and Benefits of Smart Home Automation, 2015; Nyborg/Røpke, Energy impacts of the smart home-conflicting visions, 2011; van Dam *et al.*, Do home energy management systems make sense?, 2013.
51 For dubious examples see Huhn, Smart Home, 2016.
52 Bundesministerium für Wirtschaft und Technologie, Energiekonzept für eine umweltschonende, zuverlässige und bezahlbare Energieversorgung, 2010.
53 Elsberg, *BLACKOUT: a novel*, 2017.
54 See e.g. Gößling-Reisemann, Resilience – Preparing Energy Systems for the Unexpected, 2016.
55 BBC, Ukraine power cut 'was cyber-attack', 2017; E-ISAC, Analysis of the Cyber Attack on the Ukrainian Power GridDefense Use Case, 2016.
56 World Energy Council, The road to resilience, 2016.
57 Greveler/Justus/Loehr, Forensic content detection through power consumption, 2012.
58 Shane/Rosenberg/Lehren, WikiLeaks Releases Trove of Alleged C.I.A. Hacking Documents, 2017.
59 Foucault, *Discipline and Punish*, 1977.
60 On this dystopia see Bentham/Bozovic, *The panopticon writings*, 1995.
61 See e.g. Elfvengren *et al.*, The future of decentralized energy systems, 2014; European Parliament, Decentralized Energy Systems, 2010; Vezzoli *et al.*, Distributed/ Decentralised Renewable Energy Systems, 2018; Weber/Koyama/Kraines, CO_2-emissions reduction potential and costs of a decentralized energy system for providing electricity, cooling and heating in an office-building in Tokyo, 2006.
62 See also e.g. Wang/Hao, Does Internet penetration encourage sustainable consumption?, 2018.
63 See e.g. Dusi, The Perks and Downsides of Being a Digital Prosumer, 2016.
64 Brauer *et al.*, Green By App, 2016.
65 Lin, 10 eBay Statistics You Need to Know in 2019 [Infographic], 2019.
66 eMarketer, Retail e-commerce sales worldwide from 2014 to 2023, 2019a.

67 eMarketer, Retail e-commerce sales worldwide from 2014 to 2023, 2019a.
68 Anders, Retail Industry Expects More Sales Growth In 2018, 2018; Deloitte Touche Tohmatsu Limited, Global Powers of Retailing, 2019.
69 See e.g. European Commission, Environmental potential of the collaborative economy, 2018b; Skjelvik/Erlandsen/Haavardsholm, Environmental impacts and potential of the sharing economy, 2017.
70 Gossen/Ludmann/Scholl, Sharing is caring, 2017.
71 Helbing *et al.*, Will Democracy Survive Big Data and Artificial Intelligence?, 2017.
72 Cross, *An all-consuming century*, 2000; for a very early discussion see Galbraith, *The Affluent Society*, 1958; Schor, *The Overspent American*, 1998 and many others.
73 See for an early discussion Sachs/Santarius, *Fair Future*, 2007.
74 Reisch *et al.*, Consumers in the Digital World, 2016.
75 Marshall, Google Maps Adds Location Sharing, Quietly Drools Over Your Data, 2017.
76 See also Interactive Advertising Bureau, Mobile Location Use Cases and Case Studies, 2014; Kaye, 3 Mini Case Studies Show How Location Data Is Moving Marketing, 2017.
77 See e.g. OECD, Personalised Pricing in the Digital Era, 2018; for a comprehensive review see: Van den Boer, Dynamic pricing and learning, 2015.
78 Hogan, Consumer Experience in the Retail Renaissance, 2018.
79 See on this e.g. European Commission, Consumer market study on online market segmentation through personalised pricing/offers in the European Union, 2018a.
80 See on this Chen/Mislove/Wilson, An empirical analysis of algorithmic pricing on amazon marketplace, 2016.
81 See e.g. Dahler, Do companies charge online shoppers different prices?, 2014; Mattioli, On Orbitz, Mac Users Steered to Pricier Hotels, 2012.
82 Keutel, Elektronikartikel: Wochentage mit den besten Preisen, 2015.
83 For numerous examples see also Christl, Corporate Surveillance In Everyday Life, 2017.
84 Goldman Sachs, Virtual and augmented reality, 2016; Rauschnabel/Brem/Ro, Augmented reality smart glasses, 2015.
85 Kuzyakov/Pio, Next-generation video encoding techniques for 360 video and VR, 2016.
86 Kelly, The untold story of magic leap, the world's most secretive startup, 2016.
87 Thus e.g. Wilson, The Web Is Basically One Giant Targeted Ad Now, 2017.
88 For an overview see Groß, Mobile shopping, 2015.
89 Wang/Malthouse/Krishnamurthi, On the Go, 2015.
90 See also Reisch *et al.*, Digitale Welt und Handel, 2016.
91 Wahnbaeck/Roloff, After the Binge, the Hangover, 2017.
92 Wahnbaeck/Roloff, After the Binge, the Hangover, 2017.
93 See e.g. Morgan, Will There Be A Physical Retail Store In 10–20 Years?, 2018.
94 Bundesinstitut für Bau-, Stadt- und Raumforschung, Online-Handel, 2017; Reink, "E-Commerce" und seine Auswirkungen auf die Stadtentwicklung, 2016.
95 See e.g. Ternes/Towers/Jerusel, *Konsumentenverhalten im Zeitalter der Digitalisierung*, 2015.
96 For an overview see Mangiaracina *et al.*, A review of the environmental implications of B2C e-commerce, 2015; van Loon *et al.*, A comparative analysis of carbon emissions from online retailing of fast moving consumer goods, 2015.
97 For a very optimistic view see Cairns, Delivering supermarket shopping, 2005; less optimistic are Rosqvist/Hiselius, Online shopping habits and the potential for reductions in carbon dioxide emissions from passenger transport, 2016; Siikavirta *et al.*, Effects of E-Commerce on Greenhouse Gas Emissions, 2002.
98 Mangiaracina *et al.*, A review of the environmental implications of B2C e-commerce, 2015.

99 Brown/Guiffrida, Carbon emissions comparison of last mile delivery versus customer pickup, 2014.

100 Wiese/Toporowski/Zielke, Transport-related CO_2 effects of online and brick-and-mortar shopping, 2012.

101 Asdecker, Returning mail-order goods, 2015.

102 See e.g. Grothaus, Amazon is under fire in Germany for destroying as-new and returned items, 2018; Hielscher/Goebel/Brück, Amazon destroys massive quantities of returned and as-new goods, 2018.

103 EPA, Advancing Sustainable Materials Management, 2018.

104 Eurostat Statistics Explained, Packaging waste statistics, 2019.

105 GeSI/Accenture, Smarter 2030, 2015.

106 For Germany, see Statistisches Bundesamt, Broschüre Verkehr auf einen Blick, 2013; Statistisches Bundesamt, Verkehr aktuell, 2017.

107 GeSI/Accenture, Smarter 2030, 2015, pp. 46ff.

108 Bundesministerium für Verkehr und digitale Infrastruktur, Freight Transport and Logistics Action Plan, 3rd update 2017c, p. 31.

109 Bundesministerium für Verkehr und digitale Infrastruktur, Freight Transport and Logistics Action Plan, 3rd update 2017, p. 22.

110 Bundesministerium für Verkehr und digitale Infrastruktur, Ethics Commission. Automated and Connected Driving, 2017a; European Parliament, Report on autonomous driving in European transport, 2018; Goodall, Can you program ethics into a self-driving car?, 2016.

111 Canzler/Knie, *Grüne Wege aus der Autokrise*, 2009; Canzler/Knie, *Die digitale Mobilitätsrevolution*, 2016.

112 Sachs, *Die Liebe zum Automobil*, 1984.

113 Bundesministerium für Verkehr und digitale Infrastruktur, "Eigentumsordnung" für Mobilitätsdaten?, 2017b.

114 Krzanich, Data is the New Oil in the Future of Automated Driving, 2016.

115 See e.g. Barcham, Climate and Energy Impacts of Automated Vehicles, 2014; Sokolov, Forscher, 2016; Trommer et al., Autonomous Driving, 2016.

116 Pluta, ÖPNV, 2017.

117 See Amatuni et al., Does car sharing reduce greenhouse gas emissions?, 2019; Chen/Kockelman, Carsharing's life-cycle impacts on energy use and greenhouse gas emissions, 2016; Jung/Koo, Analyzing the Effects of Car Sharing Services on the Reduction of Greenhouse Gas (GHG) Emissions, 2018.

118 See in general on the indirect effects of carsharing Evans, Cars and second order consequences, 2017.

119 See e.g. Bischoff/Maciejewski, Simulation of City-wide Replacement of Private Cars with Autonomous Taxis in Berlin, 2016; Fagnant/Kockelman, Preparing a nation for autonomous vehicles, 2015.

120 Gruel/Stanford, Assessing the Long-term Effects of Autonomous Vehicles, 2016; Milakis et al., Policy and society related implications of automated driving, 2017; Milakis et al., Development of automated vehicles in the Netherlands, 2015.

121 Gruel/Stanford, Assessing the Long-term Effects of Autonomous Vehicles, 2016.

122 For an overview of multimodal transport apps see e.g. Tracxn, Top Multimodal Transport Apps Startups, 2019.

123 Shaheen/Cohen/Martin, Smartphone App Evolution and Early Understanding from a Multimodal App User Survey, 2017.

124 Pfeiffer, The Vision of "Industrie 4.0" in the Making, 2017.

125 Bundesministerium für Wirtschaft und Energie, Industrie 4.0, 2019

126 Daugherty et al., Driving Unconventional Growth through the Industrial Internet of Things.

127 Rüßmann et al., Industry 4.0, 2015.

128 Rüßmann *et al.*, Industry 4.0, 2015; see also Roland Berger Strategy Consultants, Die digitale Transformation der Industrie, 2015.

129 Pilgrim *et al.*, The Dark Side of Digitalization, 2017.

130 ICTechX, RFID Forecasts, Players and Opportunities 2019–2029, 2019; Rand Europe/RWTH Aachen/P3 Ingenieursgesellschaft, Smart Trash, 2012.

131 Marscheider-Weidemann *et al.*, *Rohstoffe für Zukunftstechnologien 2016*, 2016; Pilgrim *et al.*, The Dark Side of Digitalization, 2017; see also Lutter *et al.*, Die Nutzung natürlicher Ressourcen, 2016.

132 Cisco, The Zettabyte Era, 2016; EMC, The digital universe of opportunities, 2014.

133 See e.g. McKinsey, The Internet of Things, 2015; Vermesan/Friess, Digitising the Industry, 2016; Zuo/Tao/Nee, An Internet of things and cloud-based approach for energy consumption evaluation and analysis for a product, 2017.

134 Andrae/Edler, On Global Electricity Usage of Communication Technology, 2015; Hazas *et al.*, Are there limits to growth in data traffic?, 2016; Lewis, Will the internet of things sacrifice or save the environment?, 2016.

135 Cisco. Cisco Visual Networking Index: Forecast and Trends, 2017–2022, 2018.

136 Bihr, *View Source Shenzhen*, 2017; ThingsCon, The State of Responsible IoT, 2017.

137 Fritzsche *et al.*, Exploring the Nexus of Digital Technologies and Mini-grids for Sustainable Energy Access, 2019.

138 Matsuda/Kimura, Usage of a digital eco-factory for sustainable manufacturing, 2015; also Hermann *et al.*, Sustainability in manufacturing and factories of the future, 2014.

139 Beier *et al.*, Sustainability Aspects of a Digitalized Industry, 2017.

140 GeSI/Accenture, Smarter 2030, 2015.

141 Bentzen, Estimating the rebound effect in US manufacturing energy consumption, 2004; Borenstein, A microeconomic framework for evaluating energy efficiency rebound and some implications, 2013; Santarius, Investigating meso-economic rebound effects, 2016a; Saunders, Historical evidence for energy efficiency rebound in 30 US sectors and a toolkit for rebound analysts, 2013; Sorrell, Jevons' Paradox revisited, 2009; Turner, Negative rebound and disinvestment effects in response to an improvement in energy efficiency in the UK economy, 2009.

142 Santarius, Investigating meso-economic rebound effects, 2016; Saunders, Historical evidence for energy efficiency rebound in 30 US sectors and a toolkit for rebound analysts, 2013.

143 Rüßmann *et al.*, Industry 4.0, 2015.

4

SOCIAL JUSTICE THROUGH AUTOMATION AND ALGORITHMS?

The year is 1830; the place, Kent, England. Some farm workers decide to take drastic action: they set fire to a threshing machine – one of those disruptive inventions, which had just come onto the market. It is the era of the early industrial revolution. In England, the enclosure movement is privatizing swathes of land that used to be available for common use. Many farm workers lose their means of earning a living in agriculture and have to find work in new industries, like textile production, instead. In this situation, the introduction of labour-saving technologies like the threshing machine poses a further threat to their livelihoods. In the years that follow, the new production technologies are subject to repeated attacks. An entire protest movement grows up, which has gone down in the history books as the "luddites". Workers struggle against the impacts of technological innovations and battle for their jobs, their wages and their social status.[1]

The early anti-technology battles during the infancy of the capitalist system teach us a great deal about the impacts of technological change, lessons that can be applied to the present day. Today, it is robots and algorithms that are changing the economy and, by extension, social conditions. Joseph Schumpeter, one of the most influential economists of the 20th century, described such processes as "creative destruction":[2] the old must yield to make way for the new to emerge. New technologies have always provoked a mixture of responses, from enthusiasm to downright hostility. The luddites were dominated by their misgivings, whereas Joseph Schumpeter was an advocate of continuous change, which he strongly associated with progress.

Digitalization is not expected to be any different: there will be winners and losers. Companies from the IT sector perceive huge opportunities in digitalization, to construct new business models and turn the "disruptive" reconfiguration of the economy to their advantage. Equally, firms working on the development of automated decision-making systems or robots can look forward to years of profit, as can the venture capitalists and shareholders in the background. Meanwhile, many established businesses dread the effect on their profits. Some lines of business look more or less doomed; take video rental stores or photo laboratories, for example. Plenty of employees in industry and service businesses worry what will become of them if their jobs are rationalized out of existence in the future. If they do not own shares in IT companies, are not trained programmers and possess none of the other skills in future demand, what are they supposed to live on? Without a doubt, digitalization touches on a range of crucial questions about economic justice.

In this chapter, we begin with an analysis of which jobs are under threat in the wake of digitalization – and how many and what kind of jobs might emerge elsewhere as a reflex. Looking back at the luddites, we ask whether digitalization might differ fundamentally from previous waves of technological innovation. Our question "Is this time different?" does not just relate to the number of jobs being made redundant or newly created. It also concerns the quality and remuneration of the new work. We will see that a feature of the digitalized world of work is ever-increasing polarization, particularly in the case of crowd working. If no changes are made to the regulatory and policy environment, and digitalization is continued on its present path, a few people will keep or obtain well-paid jobs, but many more will find themselves reduced to precarious employment. The internet is thus supporting "the return of the servant".[3] The resulting involuntary part-time labour, short-term or zero-hour contracts, internships and the multifarious nature of a gigging, entrepreneurial, freelancing and self-employed workforce is associated with a "Concierge" or "Gig Economy", which will be discussed in the following sections.[4]

Apart from labour issues, the structure of the market is decisive for the fairness of an economy. In principle, digitalization harbours great potential for reorganizing the economy to be more democratic and more decentralized. It could also reinvigorate existing concepts of a "post-growth economy" or a "distributed economy". But, at the same time, vast internet corporations have emerged and are attaining huge market power thanks to network effects and highly capitalized investors. And this economic power, in turn, translates into social and political power. In the digital economy, as it is currently developing, does the tendency towards monopolization outweigh the potential for democratization and decentralization?

In the existing regulatory environment, digitalization is also contributing to a shift in the ratio between wage incomes and investment incomes. We analyse why the share of total national income – i.e. gross domestic product – attributed to wages is falling, while the share of investment income is rising. We also show that this increasing inequality ultimately depresses purchasing power. Might this "Polarization 4.0" explain why, despite improving resource efficiency and labour productivity in the early industrialized countries, digitalization has barely contributed to growth so far?

Finally, we show that the large internet corporations are free riders on public goods because they barely contribute to the financing of digital infrastructures and other public tasks. The combination of increasing inequality, growing concentration of power and low growth has led to a new economic regime – a "digital neo-feudalism".

All in all, the signs are clear that the potential of digitalization to enhance economic justice is being under-utilized. Instead, the current prevailing trends are growing inequalities and concentration of income and power. For this precise reason, we raise one final question at the end of this chapter: do the new technologies really open up such life-enhancing and life-enriching options that they make us happy despite everything?

Jobs: is this time different?

A picture says more than a thousand words, the saying goes. In the pre-digital era when that adage was coined, photography was still a luxury. Today, it is a very different story: images are more or less omnipresent in our everyday life. Imagine that I'm standing in a shop wondering which toy to buy for my niece's birthday. 20 years ago, I would have asked the salesperson. Ten years ago, I'd have called my niece's parents on my mobile phone. Today I send them pictures of three different options and they promptly reply, telling me which gift is the most appropriate. Nowadays we use pictures to make all kinds of decisions: we check out holiday resorts and accommodation by looking at photographs or even a live video stream. Online buying decisions are based on visual impressions. Photographs are also in constant use to maintain relationships. In Instagram posts, photos are a must. And if we are interested in a person, more likely than not we will start by searching for their picture online. Today, we communicate with a thousand pictures – rather than a thousand words.

As might be expected, the number of photographs taken has grown exponentially. According to estimates, in the 1930s it amounted to one billion per year; 40 years later, in the 1970s, it had risen to ten billion, and another 40 years on, in 2012, the figure was around 380 billion photos[5] – and rising. These pictures are shared online in vast numbers: in 2016, on Facebook alone, some 350 million images were posted – daily. Added to that were another 95 million photos and videos shared on Instagram everyday, with posts garnering 4.2 billion "likes" each day – this is likely to be even higher today.[6]

Contrary to the usual economic assumptions, this rapid growth in photography astonishingly went hand in hand with an equally rapid decline in jobs. The normal rule is that, if the produced volume of a product suddenly grows, the required number of workers will grow in response. But, digitalization in the photographic industry has overturned this traditional equation. The change in the size of Kodak's worldwide workforce vividly illustrates how sweeping the consequences of digitalization can be. The transformation from analogue to digital photography meant a reduction of jobs at Kodak from 145,000 in the year 1988 to just one-tenth of that number in 2011.[7] In part, this is because digitalization has significantly improved the efficiency of the photographic process. In the 1990s, we were still taking the majority of our photos to photographic stores to be developed, whereas these are rarely found today except in virtual form. Once, an entire industry and large numbers of jobs hinged on the development of photographs; these have been lost in the process of digitalization. The other aspect is that Kodak missed out on occupying the new business segments. Instead, jobs were created by the manufacturers of smartphones and in the social media where the photos are uploaded and shared.

Fair enough, we might say: Kodak missed its moment, but jobs were created elsewhere instead. However, the ratio of jobs to revenues has fundamentally shifted in the new business segments. Back in the days when Kodak had 145,000 employees, its sales peaked at "only" US$16 billion.[8] Now, take Apple, which employed 116,000 people in 2016, a similar number – but with sales of US$216 billion.[9] So, Kodak was employing around 17 times as many people, relative to the sales figures. Meanwhile, Instagram has a total workforce of a mere 550 people, with expected sales of US$1.5 billion in 2017 from advertising on mobile devices.[10] Proportionately, that is a staggering 25 times fewer jobs than at Kodak. In sum, digitalization led to a drastic decline in the old photographic industry, while causing other jobs to be created elsewhere – but on a significantly smaller scale.

TABLE 4.1 List of industries and percentage of jobs at risk from automation in the UK

Industry	% of jobs in the industry at risk	Number of potential job losses
Chefs & catering industry	54%	2,418,763
Manufacturing	45%	1,170,000
Arts, entertainment & recreation	45%	992,350
Construction	39%	1,064,700
Wholesale and retail trade	34%	986,000
Property, housing & estate	31%	11,333
Legal profession	24%	89,947
Human health and social work	21%	376,740
Education	9%	135,900

Source: PricewaterhouseCoopers (2018), Office for National Statistics (2019).

Notes
Digitalization implies different levels of probability of rationalization for different industries and occupations. Many jobs in the retail industry, as well as low-skilled manual occupations, are especially vulnerable. Less affected are social services such as jobs in education and nursing.

The above example of the photographic industry is a spectacular demonstration of the dynamics of digitalization. The digital transformation has already swept through that industry comprehensively. Other sectors are just beginning to anticipate the disruptive changes that will ensue from digitalization. There is barely an industry that has not already undergone job losses or is not forecast to do so.[11] As an example of jobs under severe threat, take postal and delivery services: the current rapid growth in online commerce (see Chapter 3) and the corresponding rise in the number of parcel deliveries mean that the demand for parcel delivery drivers is increasing, for the time being.[12] But, in future, these jobs could be taken over by self-driving vehicles and drones, work on which is already in hand. Vehicles with self-driving features are already a reality, yet they still require a driver to steer the car manually.[13] Some 342,000 people sort mail for delivery and deliver mail on an established route by vehicle or on foot in the USA alone.[14] It is not just postal workers who are affected. Several car manufacturers are piloting fully automated vehicles as alternatives to the many lorries on American, German and British roads, which should be able to make a journey without human interaction. A scenario published by the White House in 2016 estimated that nearly 3.1 million drivers working today could have their jobs automated by autonomous vehicles in the USA. Most of the jobs are in heavy trucking. The report estimates that 80–100 per cent of these nearly 1.7 million jobs will be automated. Delivery drivers and self-employed drivers for on-demand services like Uber could face almost total automation as well.[15] These examples show that the "robot workmate" may, in fact, be unmasked as a job-thief for middle- and low-skilled workers. But, digitalization also affects those in graduate careers; take journalism, for instance. Even today, newspaper articles are being written by algorithms, particularly finance and sports news reports, which contain lots of statistics. According to "optimistic" estimates, the writing of 90 per cent of articles in future could be automated.[16] Almost no occupation will be unaffected by technological change. Routine physical and cognitive tasks will be the most vulnerable to automation. Some of the most affected jobs are those in office administration, production, transportation and food preparation. Such jobs are deemed to face "high risk" with over 70 per cent of their tasks potentially automatable.[17] All of these either involve routine, physical labour or information collection and processing activities. Digitalization of the education sector will also continue. The number of online courses and the proportion of all enrolled students who are studying online in the USA continues to rise[18] and it is easy to foresee that reductions in teaching posts could follow. In this way, robots begin to replace, not just physical tasks, but also many cognitive activities.

It seems that, whichever way we turn, digitalization is putting jobs on the line. Several studies reach a consensus that digitalization in the present regulatory environment will trigger substantial upheavals on the labour market. Just how drastic the impacts will be, however, is still a matter of controversy. Studies arrive at very different assessments of how many jobs will actually be lost. For example, a much-cited study by Carl Frey and Michael Osborne, looking at the USA, predicts that 47 per cent of jobs – across all skill levels – can be replaced by computers and robots.[19]

More recent analysis takes a more nuanced view of tasks, rather than jobs, being automated.[20] One analysis suggests that less than 10 per cent of jobs can be automated entirely and that the level of potential automation of tasks within a job varies greatly according to job and industry.[21] Recent studies take into account that most occupations are actually composed of very heterogeneous work profiles with widely varying tasks. Digitalization can usually only replace part of the range of tasks of an occupation. Another study shows a positive effect of automation on jobs in the EU. Overall, labour demand increased by 11.6 million jobs due to computerization between 1999 and 2010 in the EU 27, thus suggesting that the job-creating effect overcompensated the job-destructing effect. The study states that whilst automation decreased labour demand by 9.6 million jobs, this was compensated by product demand and spillover effects that increased labour demand by around 21 million jobs.[22] The World Economic Forum predicted that, by 2022, the total task hours completed by humans will drop by 13 per cent. Relying on job loss to automation statistics from 2018, the study suggests that 71 per cent of total task-hours are currently completed by humans, compared to 29 per cent that are done by machines. The study predicts that, if current trends continue, in just four years the average will shift to 58 per cent completed by humans and 42 per cent by machines. The use of algorithms will see an especially significant increase in specific tasks like information and data processing, while also taking over a percentage of tasks that are still performed by humans, like decision making, communicating and coordinating.[23]

All the studies agree, however, that, in principle, the bulk of tasks (meaning the individual components of the work carried out by a given employee) can be subject to rationalization. That means, if businesses successfully restructure to the point where robots and algorithms take on all tasks that can be rationalized, leaving people to do only those that cannot be rationalized, then a significant number of jobs could disappear. What seems clear is that these changes will come at costs for lower-skilled workers: work in the digital era is more connected to lifelong learning than ever before.[32] A whopping 54 per cent of all employees will require "significant" training to either upgrade their skills or acquire new skills altogether. By 2022, "everyone will need an extra 101 days of learning", according to the World Economic Forum.[33] Based on this logic, an Oxford Economics study demonstrates that machines are expected to displace about 20 million manufacturing jobs across the world over the next decade, 14 million in China alone. Each industrial robot is, on average, replacing 1.6 human workers, meaning that the number of displaced workers could reach tens of millions in the coming decade. They also suggest that low-income regions of the world are going to feel the impact of automation in manufacturing much more than average- and high-income areas.[34]

The studies presented drew spectacularly dissimilar conclusions, showing that it is far from easy to predict how many jobs are at real risk of rationalization.[24] Today, the discourse about the "future of work" appears to be arriving at a more nuanced understanding, suggesting that automation will bring neither apocalypse nor utopia regarding employment.[25]

Technological utopia: Artificial intelligence

One definition of artificial intelligence is the attempt to replicate human intelligence with the aid of computer programs.[26] The principle of "machine-based learning" is that computers are fed large volumes of data, which they analyse using self-learning techniques in order to solve complex problems.[27] The chess duels played by IBM's *Watson*, which also beat the best player in the popular American quiz show *Jeopardy*, or the victories of Google's *AlphaGo* in the Asian game *Go* are rated by some as the first evidence that artificial intelligence could even outstrip human intelligence.[28] According to forecasts, applications of artificial intelligence will make up the largest market in the digital economy in future, and this might be carved up between a relatively small number of IT corporations.[29]

What can we expect in future as the development of AI keeps advancing? Possible applications range from domestic robots and self-driving cars, through diagnostic support tools for doctors and software for emotion recognition, to applications in synthetic biology or the attempt to create a brain-computer interface. Journalism is a concrete example that highlights several risks for the economy and society even at this early stage:[30] the writing of simple articles – such as sports or stock-market news – is already being automated to some extent. Before long, we could even see automatically transcribed interviews being converted into newspaper articles with original voiceovers. Little by little, journalists could be replaced by algorithms – but what then? Will "algorithm journalism" serve as the nation's fourth estate? Or, to reframe the question: how much influence on a country's public debate and policies would be placed in the hands of the AI programmers? Ultimately, every application of AI raises such questions of sovereignty: if self-learning machines relieve us of taking actions and making decisions, then are we controlling these robots – or are the corporations behind them controlling us? And, besides: the training of machine-learning algorithms, such as deep learning or neural networks, demands tremendous amounts of energy.[31]

Apart from the question of how many jobs will disappear, for the overall employment figures what matters is how many new jobs are created elsewhere. Some studies believe that artificial intelligence, robotics and other forms of "smart automation" have the potential to bring great benefits to the economy, by boosting productivity and creating new and better products and services.[35] In an earlier study, they estimated that these technologies could contribute up to 14 per cent to global GDP by 2030, equivalent to around $15 trillion at today's values.[36] They suggest that these technologies could hold the key to reversing the slump in productivity growth seen since the global financial crisis for advanced economies like the USA, the EU and Japan. Further, any job losses from automation are supposed to be offset in the long run by new jobs created as a result of the larger and wealthier economy made possible by these new technologies. The World Economic Forum believes that emerging tech will create more jobs than it destroys, at least for the next four years. Specifically, The Future of Jobs Report 2018 predicts the loss of 75 million jobs by 2022 and the creation of 133 million jobs over the same period, for a net increase of 58 million jobs.[37] The balance between job losses and newly created jobs is central to the question: "is this time different?"

We ask this because, in earlier phases of mechanization and automation, technological innovations reduced the number of jobs per product just as they do today, but expansions in production restored the balance. And, if sweeping job losses did occur in certain economic sectors, new and previously unthought-of jobs were created in other areas. Who could have imagined in the 18th century how many factory workers would be needed after the industrial revolution? Who in the 19th century could have anticipated the vast array of service jobs that exist nowadays? So, job cuts in certain sectors were balanced out thanks to high economic growth in others. But, does this equation still hold true in the digitalized economy? Are we dealing with nothing more than the known phenomenon of structural change – just at a vastly accelerated pace? Whether it will be different this time depends on three crucial aspects.

The first is that the jobs being made redundant in the course of digitalization are different in nature from those of past eras, like those of the luddites in the 19th century. Back then, the jobs replaced by machines were largely in agriculture. Many of the labourers found work in the industrial sector, where those selfsame machines were among the products being turned out by labour-intensive manufacturing processes. Then, in the 20th century, automation led to a reduced industrial workforce while new jobs were created in the service sector. For the most part, these past waves of rationalization were geared towards replacing physical tasks. Digitalization is now changing that by providing substitutes for both physical and cognitive tasks.[38] No longer is it only the ordinary workers in industry who have to fear for their jobs; even their supervisors and parts of management are not exempted. And jobs beyond industry are also being affected, from secretaries to university lecturers. This first aspect is a sign that the wave of rationalization due to digitalization could be different from its precursors – simply because significantly more tasks can be replaced.

Second, a lot depends on the pace at which the new technologies become mainstream. This could be considerably faster than in earlier phases of industrialization. The expansion of the rail and road network took many decades, for instance, whereas new apps have far shorter development times and take a matter of seconds to download and install.[39] On the other hand, many digital functions also depend on the expansion of real infrastructures such as a high-speed data network. As we showed earlier, there are many tasks that can be rationalized, but fewer jobs that can be replaced completely. For that reason, companies intent on replacing large numbers of jobs would have to restructure themselves radically in order to be able to fire a large share of employees. This is closely linked to the question of how rapidly new technologies become mainstream; one might also say, how quickly the technology corporations manage to make the new options palatable to society and overcome barriers to adoption. Hence, the introduction of such technologies as driverless cars or delivery drones will stand or fall on whether laws are changed, moral attitudes adjusted and new infrastructures built. How quickly digitalization puts pressure on the labour market will not be determined by its technological feasibility alone, but rather by the policy and regulatory environment and by acceptance within society.

A third aspect to the question of whether digitalization will lead to a net loss of jobs is the level of economic growth. This is because, the greater the number of jobs rationalized due to digitalization, the more jobs would need to be created elsewhere. Even if we have no idea at all what kind of jobs these might be, for the moment, one thing is clear: the more comprehensive the rationalization, the higher the rate of economic growth needed to create enough new jobs. The growth rates in highly industrialized countries are lower today than in earlier phases of radical technological innovation.[40] The World Economic report referred to above, which forecasts an overall gain of jobs, is based on trends and assumptions emerging from the collective actions and investment decisions taken by companies and governments and that employees also stand to benefit by virtue of rising wages.[41] Investments and rising purchasing power stimulate economic growth, the argument goes. However, it is doubtful whether this assumption is realistic under the current macroeconomic conditions. As we will show later on, digitalization is rather helping to keep growth rates low, which may well leave more people unemployed or in low paid jobs.

From the perspective of global justice, an additional important question is which countries the new jobs are generated in. In the past, many countries, in particular in Asia, have followed an export-led manufacturing route to rapid growth. Automation has already eroded this pathway and can be expected to continue to do so, as the advantages of abundant cheap labour are worn away and "re-shoring" production closer to rich country markets increases.[42] In many sectors in developing countries, like agriculture, smallholders may lose out under increasingly automated processes.[43] Again, digital technology will mostly benefit skilled workers at the expense of those less skilled and these tend to be more concentrated in richer countries. Of course, the export-led growth strategies in the past have also been accompanied by many social and environmental problems. Whether the prospects of sustainable development in the global South have improved or worsened due to digitalization is still to be seen. For the time being, we can draw an interim conclusion that digital rationalization can lead to the replacement of significant parts of the workforce with robots and machines. The level of growth required to generate equivalent numbers of new jobs is not only unknowable, but would also be environmentally unsustainable, as we have shown at length in Chapter 3. The loss of jobs has far-reaching implications for questions of economic justice. If people lose their jobs, they first have to deal with depletion of their financial resources. We will turn to the income effects of digitalization in the next section. What is more, unemployed people simultaneously lose contact with their colleagues, often one of a person's most important social networks. A further issue is that, in our society, a (good) job is a symbol of social participation and prestige. Reflecting all this, empirical studies show that unemployment almost always has a devastating effect on people's satisfaction with their lives.[44]

But, could this not also be turned to good purpose? Might we gain more time for family, hobbies, art, education, social engagement and other activities unrelated to paid work? Might we bring in shorter working hours for the many, rather than consigning half of the people to unemployment so that the other half can have full-time jobs? Might shorter working hours for men contribute to a fairer distribution of caring and reproductive work between the genders? Might digitalization present us with a huge "gift of time" and hence an important contribution to a social-ecological transformation of society? Could we use digitalization to become less dependent on economic growth?[45]

Sadly, the lesson from history is that there is no automatic mechanism for making use of new technologies to reduce working hours. In the past 250 years of industrialization, shorter working hours have always been accomplished through social power struggles, often led by trade unions. But, today more than ever before, employees of almost all skill levels are up against international competition, which weakens their negotiating position vis à vis employers. And digitalization cranks up power disparities even further, as we will show in the following sections. Overall, making positive use of rising work productivity to give people more leisure time and distribute work more equitably seems a remote prospect.

To sum up: digitalization will bring about a rationalization of work and severe upheavals on the labour market can be anticipated. The speed and dimensions of these, however, do not depend primarily on what is feasible technologically, but on what is achievable societally. Potentially the benefits of rising work productivity could be used to free up time for care work and leisure – and thus become part of a social-ecological transformation. But, no such development seems to be on the horizon so far. Meanwhile, relying on growth to soak up redundant labour capacity as it did in the past seems neither realistic nor environmentally desirable. However, if digital options could be implemented rapidly, with low growth and no redistribution of work, this time really could be different: significantly more well-paid jobs are being lost than are emerging elsewhere. This is forcing people to accept poorer paid and precarious forms of employment, as we will see in the following section.

Return of the servants

Digitalization presents opportunities to do more than merely make people's lives more pleasant thanks to rising work productivity. Positive visions also emphasise that boring routine functions would be replaced, so that people can focus on challenging and creative tasks and the world of work can be shaped more flexibly.[46] Journalists could concentrate on background analyses and opinion pieces, while simple texts are taken care of by algorithms. Another possibility is that unpleasant and physically strenuous jobs might disappear, which would improve the well-being of working people.[47]

How desirable this is, from the viewpoint of a social-ecological transformation, is a contentious matter of debate. To state the two extremes: on the one hand, there are progress optimists who think that entrusting work to robots is the way to liberate people from the yoke of work and give them time to devote to more worthwhile activities.[48] The opposing perspective is that work is part and parcel of a good life, and that the production of objects and all kinds of necessities of life is something from which people derive enjoyment and a sense of achievement and meaning. In our view, a golden middle path makes sense: there is no reason not to turn over a share of the work to machines. But, a far more exciting question is how they can be used to make work a humane and meaningful activity. Digitalization should therefore not be geared solely towards productivity gains, but also towards improving working conditions and making sufficient human resources and remuneration available for caring and reproductive work.

How digitalization can be used to improve work situations is a question that ties in closely with the environmental aspects covered in the previous chapter. Under a rigorous system geared towards environmental sustainability, digitalization would not threaten as many jobs – because digitalization itself is still always based on the use of natural resources and energy. We know from environmental economics that how much a particular input is used depends on its price in relation to other inputs. If energy and resource consumption – and hence digitalization – became more expensive, then very different production conditions would arise, setting the scene for a different, lower-key approach to digitalization and requiring the retention of a larger workforce.

The way in which digitalization is currently proceeding does not measure up to any of the three paths outlined. As we will see below, it is not relieving large numbers of people of the need to work or significantly reducing their workloads; nor is it facilitating good, humane working conditions across the board. Instead, it looks likely to lead to polarization, both on the labour market and in incomes.

As a graphic example of developments in the labour market, let us take the phenomenon of crowd working. What do a software developer, a skilled craftsperson and a cleaner have in common? As of recently, all of them can offer their labour via the internet as part of a large "crowd" of workers. Not so long ago, software developers were hired by companies as permanent staff. Today, more and more of them apply online for short-term contracts, which they perform as independent contractors. Likewise, skilled craftspeople always used to be permanently employed in firms, but the internet is making it much easier to hire an individual craftsperson directly instead of contacting a firm. And, although cleaners were often self-employed before, today they can increasingly be booked over the internet. What they all have in common is that they work on a self-employed, fee-earning basis and therefore do not have insurable jobs with social protection, job security, paid holiday entitlements and so on. Via the internet, they are in competition with many others, and the growing use of online ratings intensifies the competition between them even more.

Generally speaking, "crowd working" is understood to mean the decentralized organization of workers via the internet. Potential suppliers (for example, a craftsperson) and potential customers (for example, an apartment owner) are brought into contact via a platform (for example, Upwork).[49] The contracts are typically concluded directly between suppliers and customers, and frequently include conditions specified by the platform – which vary greatly between different operators.[50] The number of people employed in crowd working is currently soaring. According to a survey, 9 per cent of the population in the UK and the Netherlands, 10 per cent in Sweden, 12 per cent in Germany and 19 per cent in Austria have engaged in crowd working – with only minor differences in the responses of women and men. In most cases, these are side jobs. Nevertheless, crowd working represents the primary source of income for 2.4 per cent of respondents in Austria, 2.6 per cent in Germany, 1.7 per cent in the Netherlands and 2.8 per cent each in the UK and Sweden.[51] The authors also suggest that many people choose crowd work because they have no other income stream, rather than as an active career choice. The majority of crowd workers engage in multiple types of crowd work, rather than specializing in a single form, and are actively seeking more regular types of work.

Crowd working takes very different forms. First, there are activities that are carried out online in the metaphorical "cloud" (also called "cloud work"). Second, crowd working also includes services that are brokered online but completed offline ("gig work"). In crowd working, the degree of qualification and the volume of work involved can vary enormously. So, there are highly demanding jobs such as programming work or layout assignments; but, there are also simple tasks that can be completed on a massive scale. One job on the Amazon Mechanical Turk platform involves nothing more than deciding whether two Amazon shop links actually show the same product or item. Another typical content moderation HIT ("Human Intelligence Task") on the platform includes viewing photos, video or text and marking offensive or sensitive content.[52] Examples of simple gig work coordinated via platforms are household services such as repairs, cleaning services or driving and delivery services.

Descriptions of crowd workers range between the extremes of "autonomous, flexible and free" and "underpaid, uninsured and insecure". The positive vision of the crowd worker is to be able to work wherever, whenever and as much as one likes. How about living in a Caribbean beach house? Going for a morning swim, doing a few hours work on the computer under a shady palm tree and surfing in the afternoon? Along similar lines, incidentally, Karl Marx had predicted that "his" classless society of communism would make it possible "to do one thing today and another tomorrow, to hunt in the morning, fish in the afternoon, rear cattle in the evening, criticise after dinner, just as I have a mind".[53] The fact that crowd working is only a side income for many is certainly part of the explanation.[54] But, large differentials in pay are also a factor as those crowd workers who lack full-time employment are scrambling to live off of their gigs. In an empirical investigation of the labour market effects of crowd work in the USA and EU, the effect of crowd work on earnings is negative and significant. The findings from a number of online platforms indicate that, overall, crowd workers earn about 70.6 per cent less than "traditional workers" with comparable ability, while working only a few hours less per week. These results hold both for European and American platforms.[55] It confirms the results from descriptive analyses of previous studies where earnings from micro-tasks are well below national averages.[56] Most crowd workers would prefer permanent employment over contract work.[57] The terms and conditions in some areas of the market are appalling. On many of the platforms, contracts stipulate that customers can withhold payment if they are not pleased with the work. And, since work in the gig economy does not come with benefits, crowd workers must cover their retirement, health care and operational costs out of this income. Crowd workers often work below market rate, or even for free, in order to have more gigs and ratings on one's professional profile. This gives huge power to the employers, while consigning workers to the role of supplicants. Recently, unions and worker organizations have started to defend the rights of crowd and platform workers and have taken action to improve working conditions in several countries.[58] But, the existing insecure working conditions, poor social protection and low hourly rates of pay are clear signs that a large proportion of crowd workers are the new day-labourers of the digital society.

Do crowd working and the new digital options in other occupations provide opportunities for more gender equity? At first glance, this is definitely the case: it would be a way of making family and work more compatible, perhaps by organizing working times more flexibly and enabling people to work in home offices. At the same time, there is a risk of exclusion, particularly for women, who on average still undertake the lion's share of care work even today. Some studies argue that platform work has no intrinsic effect on gender-dependent outcomes, arguing that crowd work arrangements do not tend to reinforce discriminations based on the sex of the worker, due to the relative anonymity that service providers enjoy on the platform. Clients are, indeed, often unable to ascertain the gender of online service providers.[59] However, the analysis of the gender pay gap among crowd workers reveals that women earn an average 82 per cent of what men earn.[60] One explanation of this gap is women's domestic responsibilities.[61] But, if digitalization means that people are, in effect, always working – even in the evenings when the children are finally in bed – then little has been gained.[62] More equal distribution of work – including shorter working hours and more reproductive work for men – is, therefore, an important demand, as we will elaborate further in Chapter 6.

And then there is the international perspective: the digital devices and infrastructures that are essential for remote working are based on environmentally questionable exploitation of resources and are often produced in inhumane working conditions, a point we dealt with in Chapter 3. While on the face of it, the manufacturing of billions of digital devices may mean jobs for many people, particularly in the global South, it is nevertheless worth asking: does digitalization improve the quality of their working lives? So far at least, the "brave new world of work" has not quite reached the approximately 1.3 million workers at the Chinese electronics corporation Foxconn, who manufacture the hardware for our smartphones, tablets and computing centres,[63] to cite just one example. How about pregnant women, who are often dismissed without notice because they are no longer 100 per cent available? Can they assemble the microprocessors at home by telework? It is evident from the environmental and social drawbacks that digital lifestyles are often part of an "imperialist mode of living", where comfort of some is associated with the exploitation of others.[64]

Another instance of extreme inequality internationally is the distribution of access to and use of digital devices and the internet (see Figure 4.1. Although rich elites even in the poorest countries have long had smartphones, Wi-Fi and more, there is still a dramatic "digital divide", a huge disparity in the distribution of the fruits of digitalization.[65] Many regions of the world still have patchy access, if any, to digital networks, which is not really surprising, given the prices of smartphones and other digital devices. Even in the countries of the global South, these are already prevalent among the urban middle and upper classes, but, for two to three billion of the world's people, it is simply unaffordable to benefit from digitalization for the time being.

The issue of data protection also has an important bearing on crowd working. In the USA, for instance, female Uber drivers have been harassed by

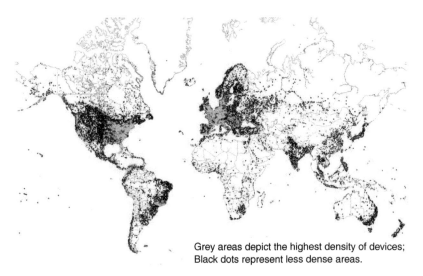

Grey areas depict the highest density of devices;
Black dots represent less dense areas.

By far the greatest number and density of devices connected to the Internet are located
in the "old" industrial North: Europe, North America, Japan, Australia/New Zealand. Some
regions of the Global South have done much to open up access; for example, regions in
Brazil, China, India and Mexico. Large parts of the rural regions of Latin America, Asia and
especially Africa are almost excluded from the virtual world.

FIGURE 4.1   Digital divide: the global distribution of networked devices

Source: own presentation after Hines (2016), based on www.shodan.io.

passengers after having used their service. Using a combination of the Uber "lost
and found" system and the Apple "Find my iPhone" app, passengers were able
to find out women drivers' addresses and pay them a call at home.[66] Furthermore, crowd-working platforms make use of digital tools to monitor their staff
to varying degrees.[67] Covert screenshots can be made or keystrokes and mouse
movements recorded. Overall, the constant process of appraisal and evaluation
of the crowd workers generates a level of pressure that is of such magnitude it is
completely out of sync with the activity or task.[68] Crowd work is often characterized by a high level of surveillance.[69] Crowd workers become functionaries in an
algorithmically-mediated work environment and are being objectified.[70] Hence,
it is not uncommon for crowd workers to be treated as "glass day-labourers".

The labour market as a whole comes under similar influences from digitalization,
as is the case in the crowd-working sector: on one side of the equation are the few
workers who are in high demand, often for their IT and other technical skills. They
enjoy generous pay and good working conditions. On the other side are the many
who have to hire themselves out precariously for low pay with poor to non-existent
social protection. This polarization of the labour market is based on two phenomena.

First, lower-income jobs are harder hit by digitalization than those paying high
incomes. Consequently, it is the low and medium earners who will lose their

employment: the probability of being affected by rationalization correlates with pay level.

Second, new jobs will arise in the very well-paid segment, but very few at the middle income level. Let us take another look at the examples from earlier of where jobs are being replaced and with what: in logistics, with self-driving cars and drones; in journalism, with "bots"; in the financial markets, with algorithm-based trading; in industry, with robots and so on. These examples have one thing in common: an extra need for a particular occupational group so that the algorithms and robots can unfold their artificial intelligence. For the foreseeable future, software and robots are developed by people – by hardware and software developers. Workers need to be better at complex problem-solving, teamwork and adaptability.[71] Forecasts predict that, by 2022, there will already be a significant shortage of skilled people in these areas – engineers, natural scientists and mathematicians, data analysts, e-commerce and social media specialists.[72] But, as we showed in the previous section, not enough well-paid jobs will be created in these occupations to make up for the jobs lost elsewhere due to digitalization. That is why so many jobs in the low pay segment are being created today and the trend is only likely to intensify in future.

The author Christoph Bartmann describes the direction in which we are being taken by the new division of labour in the digital age as "the return of the servant".[73] In his book on the subject, he uses the example of a typical upper middle-class household in New York City. This means households in which both partners are employed in well-paid positions. Often, they live in apartment buildings with concierges, who take care of every possible domestic

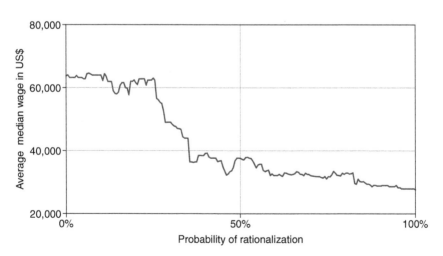

With rising income there is a decreasing probability that the job can be rationalized due to digitalization.

FIGURE 4.2 Job losses by income levels

Source: Frey and Osborne (2013).

service, from supervising workmen to taking delivery of parcels. In addition, many residents hire cleaning or housekeeping help, such as nannies for their children, while Uber drivers fulfil the function of chauffeur. In this way, they seek to externalize as much reproductive and care work as possible.[74] Many of the services that households call upon for this purpose are transacted via websites and apps. Of course, wealthy households have had nannies and housekeepers for centuries. But, the recent resurgence in the phenomenon is due to digitalization. The precariat has reached a significant part of the population. In 2018, the Bureau of Labor Statistics estimated that about 13.9 per cent of the workforce in the USA had either "alternative arrangements for their primary employment" or reported to have temporary jobs deemed to belong to the growing precariat.[75]

And, so, the story comes full circle: precisely here, jobs are materializing that digitalization and other factors are making "redundant" elsewhere. Unlike in the past, when a low-skilled young person could still find a permanent job in industry or a skilled craft, in future many of them will have to offer themselves for hire via online platforms as self-employed taxi drivers, domestic helps or jacks of all trades. These people have very little choice but to take poorly paid, insecure and precarious jobs, because permanent and better-paid employment is no longer available. From a gender perspective, too, this development is problematic. Traditionally, more men than women learn the skills that industry will demand and remunerate well in future – natural and engineering sciences, software development, etc. – and this is no less true today.[76] Thus, the "gender pay gap" could actually grow rather than diminish.

Overall, the picture that emerges on the labour market is this: jobs at all skill levels are being rationalized, but low-skilled workers are worse affected than the better qualified. Some of those made redundant will find work in the highly skilled segment, where new jobs are being created. Underqualified workers will have no other choice than hope for possibilities to train and qualify for higher-skilled job opportunities. However, many will have no choice but to accept low paid and precarious jobs if they are to earn a living at all. Both developments are two sides of the same coin. The well-paid specialists and managers, those from the IT sector for instance, have plenty of cash to outsource many of their domestic chores. However, this entire analysis is based on the assumption that existing trends will continue in the future, because the economic policy environment does not change significantly. But, could digitalization not be put to quite different uses as well? In the next section, we discuss how far digitalization could transform the economy more fundamentally in the direction of fairer and, especially, more democratic economic structures.

The opportunity for economic democracy

In 1968, Stewart Brand, who was later to become a famous border-crosser between the alternative scene and Silicon Valley, set off on an adventurous journey with his wife Lois. Driving their pickup, they did the rounds of California's education fairs, taking along their newest "baby", the earliest issues of the Whole Earth Catalog. The book was a cross between a manual and a discussion platform for the "back-to-the-land" movement of the hippie generation of 1968. The catalogue aimed to give people all the information they needed to live an autonomous, alternative life. 37 years later, in 2005, Steve Jobs described this catalogue as a kind of ana-logue precursor of the internet and one of the great inspirations for the digital innovations of the last few decades. The Whole Earth Catalog is symbolic of the hopes that were vested in the internet. Its most important aim was to contribute to a radical democratization process.[77]

Whether digitalization on a political level strengthens or weakens democracy has been a matter of debate for some years. Democracy is not simply about mark-ing a ballot paper with a cross every few years. As sociologist Jürgen Habermas persuasively argued, the quality of public discourse is key to the quality of democ-racy.[78] How is public debate conducted and, above all, who is permitted to take part? For many years, chat rooms, blogs, discussion forums and social media were considered highly promising tools for shaping a more democratic public debate – on the strength of three attributes in particular: first, the internet can be used to distribute information rapidly over long distances so that the population is always kept up to date. Second, communicating with one another via the internet costs comparatively little, which means that almost everyone can join in. Third, on the internet everyone can communicate with everyone else and, what is more, in any direction, whereas communication in the past, especially via the mass media, was understood in only one direction: from sender to recipient.[79]

Particularly in the last few years, it has become evident, however, that three developments stand in the way of these opportunities to improve public dis-course. First of all, since Edward Snowden's revelations, it has been known that security services are covertly monitoring the entire internet and gathering vast quantities of confidential information about very large numbers of people. But, a private sphere, i.e. a protected space in which individuals can think and do as they please, is a key prerequisite if citizens are to be able to take action within a democracy. Total transparency is the arch enemy of democracy. A second prob-lem is the question of whose inputs and opinions are actually heard on the inter-net and whose are not. This is increasingly influenced by social media. More and more people read the news on platforms like Twitter and Facebook. But, the underlying algorithms that select what appears in newsfeeds are constructed by a few profit-motivated companies. This puts them in possession of vast social power, because whoever controls the information can influence the discourse. Third, it is necessary to question how far communication in the digital space can really be understood as a common, public discourse, because, when people go online, it is as if each of us is surrounded by a filter bubble. We read what our own friends write and what the social media algorithms recommend to us on the basis of our past reading behaviour.[80] If everyone perceives a truth of their own, though, this can hardly result in a common public discourse. And it becomes difficult to find common solutions. In the words of Jürgen Habermas: "For the time being, virtual space lacks the functional equivalents of those structures of public discourse which collect up the decentralized messages and select and synthesize them in revised form".[81]

It is, therefore, questionable how far the internet has fulfilled any hopes that it might substantially extend democracy on the political level. In fact, a growing number of studies lately have been highlighting sizeable risks to democracy. Granted, the availability of information has improved and people can exchange views in forums, comments columns, tweets and posts. Nevertheless, the discursive power is in the hands of large commercial players. Furthermore, there has been a quantum increase in surveillance options. This prompted cultural researcher Harald Welzer to talk about a "smart dictatorship".[82] And Jaron Lanier, a Silicon Valley scientist-pioneer, alerted us to the dangers of social media and explains why its toxic effects are at the heart of its design: "Our early libertarian idealism resulted in gargantuan, global data monopsonies."[83] In his book, *Ten Arguments For Deleting Your Social Media Accounts Right Now*, Lanier draws on his insider's expertise to explain precisely how social media works by deploying constant surveillance and subconscious manipulation of its users: "We're being hypnotized little by little by technicians we can't see, for purposes we don't know. We're all lab animals now."[84]

Less widely discussed by far is the question of whether digital technologies have the potential to bring about democratization of the economy. Some of these potentials tie in with aspects discussed in Chapter 3. There we saw that a decentralized energy system could be more resilient and environmentally sound by design. However, decentralization of the energy system also implies that the wind turbines, solar panels, small hydroelectric power station or biogas plants should be owned by many thousands of private individuals, cooperatives or municipalities, and not by a handful of huge energy corporations. That would be more democratic because power over the energy supply arrangements would then be distributed among a large number of stakeholders. It would also be more equitable because the financial gains would go to numerous people instead of comparatively few owners.

Likewise, the phenomena of "prosuming" and "sharing" addressed in Chapter 3 have the potential to shape economic processes more democratically: by signing up for car pooling, many individuals could travel more cheaply or make their own journey more affordable. Collaborative platforms promoting local production have also emerged in the food industry. If platforms like the French *Agrilocal* or the German *Marktschwärmer* try to bring together local suppliers and public buyers, it supports decentralized agriculture.[85] Another type of platform only involves private individuals, who, according to a peer-to-peer logic, are both suppliers and users. People organize to exchange fruits, vegetables, fish, meat, eggs, mushrooms, seeds, plants, honey, pasta and spices via platforms.

There are many examples in the USA such as *LA Food Swap* in Los Angeles and *Chicago Food Swap* in Chicago.[86] What matters most is who controls the platforms. When Amazon was launched, people raved about its usefulness to small publishers for distributing their books. Today, Amazon has colossal market power and profits disproportionately from the added value generated. Similar trends can be observed with regard to Uber, Airbnb or commercial carsharing providers.

New formats, known as platform cooperatives, offer a way of jump-starting the democratic potential of sharing and prosuming. At the user interface, these are internet platforms, like eBay or Airbnb, but they are held in collective ownership by cooperatives, municipalities or other community organizations.[87] In the USA, there are several taxi cooperatives, such as Trans Union Car Service or Union Taxi, providing local alternatives to Uber. Fairmondo is trying to establish an alternative to Amazon, Stocksy is a stock photo site where contributing photographers are also owners and Backfeed is a platform to create platform cooperatives powered by blockchain technology. Backfeed understands its services as a "Social Operating System for Decentralized Organizations enabling massive open-source collaboration without central coordination".[88] We describe other examples in Chapter 6. What makes these alternatives more democratic is that, not only are the means of production – software, servers and distribution channels – in shared ownership, but decisions on fundamental strategy and management are made collectively by the providers, users and other stakeholders.

Another and older success story about how digitalization can be used for a decentralized reform of the economy is the open-source community. In 1985, US activist and programmer Richard Stallman published one of the seminal documents on which the movement was founded. Stallman argued that software programs are common goods since everyone can benefit from them. Codes and programs should be freely accessible and unrestricted because this is the form in which they can contribute to the greatest social good.[89] A full decade earlier, Bill Gates wrote an open letter demanding that software had to be paid for so that programmers could be rewarded for their work.[90] That marked the birth of the two opposing paradigms of digitalization: open versus closed, or – to put it another way – collective and bottom-up versus proprietary and top-down.

Today, these two paradigms are still in competition with each other, particularly when it comes to software. The many Apple and Microsoft products represent economically successful closed systems. But open-source projects have also produced some success stories: nowadays, the world's most important library is no longer a large building in London, but the internet site Wikipedia. It is freely accessible, anyone can help to enlarge and improve it, and it is operated by a non-profit organization. And, three decades after the invention of Windows, open-source operating systems are still alive and well, the best-known being Linux. Linux is even more successful in cloud systems and the internet of things, where it is implemented in over 80 per cent of devices.[91] Another long-running success story was the open-source internet browser Firefox, although it has lost market share in recent years.[92] In view of the massive financial resources that commercial software corporations invest in closed systems, it is astonishing how open source goes on and on punching above its weight. What is its secret? Open-source projects are so successful because they utilize the greatest strengths of the internet: knowledge-sharing combined with access for all interested developers and users.

The countless sharing, prosuming and open-source products and projects show how digitalization can contribute to shaping a fairer and more democratic economy. But, can the digital options also be used to bring about changes in the entire economic system, to transform existing private sector capitalism into a cooperative and democratic economy?

Let us briefly revisit the 1970s once again. The concept of sustainable development in its current meaning had yet to become established, but there was much discussion about how to pilot the countries of both the global South and the global North onto a sustainable course. In the countries of the South, the question was how they could emancipate themselves from neocolonial structures; in the countries of the North, it was how to solve the growing problems of industrialized society – first and foremost environmental degradation, but also people's increasing social isolation and the unanswered question of how a good life might look in the post-production-line era. One of the most influential books arising from these discussions was "Small Is Beautiful", published in 1973 by economist Ernst Friedrich Schumacher.[93] His main thesis was that small production units are better, for nature conservation and for social harmony. He argued, for instance, that decentralized, small-scale food production is not only more environmentally sustainable, but it also distributes income and influence more fairly. Added to that, it enables meaningful work, unlike the industrial-scale agriculture that was coming into its own at the time.

Building on these arguments, different models have been developed in recent decades to describe the shape that re-regionalization and decentralization of the economy might take.[94] Equally, recent "post-growth economy" models contain proposals that can be applied to a social-ecological style of digitalization: ideas such as regionalization, repairing and prolonging the useful life of products, and the use of productivity gains for more leisure time and social activities.[95] The idea is not to convert the economy in its entirety to subsistence or regional production, but rather to increase the weight of these elements while another component of production (a lesser share than today) is still based on the international division of work. This transformation can be described in terms of the "economic subsidiarity" model: as much local production as possible, as little supplementary global market production as necessary.[96] Proponents of the "distributed economy" argue along similar lines. Again, this entails increasing the proportion of local production while retaining global value chains, but on a smaller scale than today. Three archetypes of economies can be defined: a "centralized" system in which the structures are grouped around one central point; a "decentralized" system in which many different centres exist, which are only marginally linked

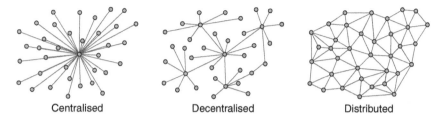

| Centralised | Decentralised | Distributed |

In a centralized economy, products travel from the suppliers to a central location, from which they are distributed to consumers. In a decentralized economy this happens across multiple locations. In a decentralized-networked economy, products are traded on different channels between multiple actors.

FIGURE 4.3   Centralized, decentralized and decentralized-networked economies

Source: Wikipedia (2017).

with each other; and a "decentralized-distributed" system that is characterized by highly interwoven structures.[97] The latter two types would go hand in hand with re-regionalization because the geographical distances would diminish.

Digital technologies – applied in the right way – could help to overcome many of the old problems of decentralized economic models. The reality is that, despite being organized on more environmentally sustainable, equitable and democratic principles, regional and decentralized modes of production and consumption still face some crucial reservations. A common fear is that we could be thrown back to the era of the village – with inefficient modes of production and conservative structures, cut off from the rest of the world. This is exactly where digitalization can help to unite the best of both worlds: environmental sustainability and a high degree of local community can be combined with a modern-day free society, material prosperity and global connectivity. In other words, the internet could be used to realize the option of "cosmopolitan local-ism". For, with the help of digitalization, we can push back the boundaries of what is feasible on the local level.

In some sectors, such as food production, a decentralized and regional produc-tion structure is still easy to imagine even without digitalization. Fertile land is still available in many places on Earth and the necessary tools and know-how for small-scale, agro-ecological farming systems are comparatively easily available and often "low-tech".[98] It is a different matter in other economic sectors. To manu-facture trains, electrical devices or mass-produced items like tools or kitchen utensils on a small scale would be either too inefficient or plain impracticable for technical reasons. But, digitalization opens up new options for decentralized pro-duction in different economic segments. scientists scattered all over the world can cooperate unproblematically. Software developers can design programs decen-trally and musicians can compose intercontinentally nowadays. In short, when the goods being produced are nonmaterial and can be digitally transmitted, there are almost no limits to the decentralized structure of the economy. In the local production of physical goods, the situation is different by nature. But, even here, digitalization can contribute to more decentralized modes of production; namely by opening up the following four potentials: information flow, access, coordination and new production technologies.

First, everyone with an internet connection has unlimited access to global information – and, with the help of search engines, can instantly find whatever know-how they need. There is no lack of guides on vegetable-growing, needlework or the carpentry skills to build a table. Even instructions on how to brew local beer or how to make shoes – not just for personal consumption, but for a local enterprise – are just a few clicks away. Repairs are a particularly good example of how new digital options can be used – in this case, with the important effect that products can be used for longer, making for a more sustainable lifestyle. The internet offers countless guides on repairing appliances like refrigerators and toasters, but also electronic devices like computers and smartphones. This free pool of knowledge can be drawn on either for personal needs or as a basis for local service businesses.

Next to information flow, "access" is the second key factor, meaning access to both products and business markets. The know-how to produce or repair something oneself and to turn that into a new business is only useful if all the necessary production inputs can be obtained and there is some means of selling the manufactured products or offering one's repair services. If someone wants to open a business that specializes in repairing mobile phones, they need specific spare parts. If someone wants to start manufacturing cargo bikes and the local market is insufficient to support turning it into a business model, then customers from a wider region are needed – and they can be reached via the internet.

Third, digital services enable us to cut out the middleman. Whether manufacturers were producing long-life food products, electrical appliances, furniture or clothing, they used to rely on cooperation with a grocery, electrical retailer, furniture store or boutique. Today, they can sell the products via their own website or via platforms with relatively little effort. In fact, platforms such as Amazon are increasingly positioning themselves as new middlemen, skimming off some of the revenue. But, this need not be the case. As an alternative, multiple merchants can organize themselves into cooperatives, as happened in the case of Fairmondo, which originated from Germany and has expanded to the UK (see Chapter 6). Blockchain technologies are another, new strategy for avoiding middlemen on the internet. The most famous example is the cryptocurrency Bitcoin. Blockchains establish a decentralized accounting system that requires no central intermediary and simultaneously prevents any individual player from tampering with contracts or agreements. These approaches – the technicalities of which are not that easy to figure out – are based on encrypting bilaterally concluded contracts between suppliers and users and linking them with the contracts of others, so that agreements cannot be manipulated. For example, tenants of apartments could enter into contracts directly with interim subtenants without having to pay commission to a middleman such as Airbnb. In this way, blockchains can solve a longstanding problem of decentralized organization: how to coordinate many actions without centralized coordination. Nevertheless, it is doubtful whether blockchain applications, which are currently still extremely energy intensive, can be designed to be viable from an environmental standpoint (see the text box on "Blockchain" in the section "The reality of monopolistic power?").

Fourth, new production technologies combined with digital transmission of information make decentralized organization easier. In the previous chapter, we already discussed the options for a decentralized energy supply. These may work out if new technologies such as photovoltaic cells or biogas plants are combined with a new grid infrastructure that is made possible by digitalization. 3D printers

Technological utopia: 3D printers

In 3D printing, it is not pages of text but whole objects that are "printed". A computer file contains the information about an object, which is then created by the printer using additive processes.[101] Over 100 materials are already printable. Car parts and furniture are among the things that have been printed using industrial printers. Home users can already print out small replacement items such as lids or switches; looking forward, it will also be possible to print more complex components, e.g. for electronic devices. This way, 3D printers may encourage people to carry out repairs themselves and practise urban self-sufficiency.[102] What is more, 3D printing is widely seen as offering great potential to decentralize the economy: if lots of the goods we need can be manufactured locally, then central mass production will be just as superfluous as the polluting transportation of goods.[103]

As yet, hardly anyone has a 3D printer at home. But, some hobbyists and do-it-yourself enthusiasts use them in "makerspaces", "fab labs" or other open workshops. That said, there has been a certain amount of criticism that the objects printed there are not always especially useful,[104] in that they are rather more likely to be gimmicks like plastic dinosaurs or other unnecessary items in the "plastic stuff" category. It is also doubtful whether more complex products can really match the quality of industrially manufactured goods and whether this form of production is not, per item, even more resource and energy intensive.[105] And it stands to reason that, if we can print any imaginable item, overall consumption could increase. Will there then be a surge in global demand for material for these printers? Some experts hope that, in the long term, 3D printers could use plastic waste as a raw material.[106] But, which materials will be used to print complex parts? How much power will be hogged by printing, and what will happen to the waste?

are a second example: these make it possible to manufacture small quantities of products for which mass production used to be the only cost-effective option – particularly products and spare parts made from plastic and other useable materials (metals, ceramics, concrete, tissue cells etc.).[99] And many of the technologies that we introduced in the section "Jobs: is this time different?" could be of use for a more regional economy. Until now, one of the main arguments against regional production was the claim that it was too inefficient. Digitalization now makes it possible to organize local production considerably more productively. For instance, efficiency gains aided by precision farming can also be utilized for urban and regional food production that is environmentally sound and productive at the same time.[100] And the ongoing efficiency improvements in logistics and delivery services could help enhance the efficiency of access to local production.

Digitalization has the potential to be part of a mix that was almost inconceivable in the past: a modern, cosmopolitan way of life, yet with decentralized economic structures. The vital ingredients come from two different scenes: from the computer scene, which proclaims the benefits of new technologies and the open-source philosophy; and the alternative scene, which brings to the table a political mindset geared towards solving environmental and social challenges. This mixture calls to mind the counterculture movement of Silicon Valley, which we described in the introduction to this book. But, considering all these opportunities, does it not give us pause for thought that, for the past few years, we have mainly witnessed the rise of gigantic commercial internet corporations with a propensity for monopolization? And that the headlines about successful start-ups from Silicon Valley tend to discuss billion dollar deals rather than, say, ways of bringing new life to more remote regions? Obviously, there are influencing factors at work that are thwarting a gradual transformation of our economy towards a decentralized economic democracy, and which have instead turned the digital economy into the spearhead of capitalism. Why is that?

The reality of monopolistic power

The real trend today is heading far stronger in the direction of monopolization than of decentralization. Commercial internet giants are carving up key functions of the internet among themselves. These are publicly listed corporations with branches all over the world. The first sign of the power wielded by these IT and internet giants is their stock market value. Seven of the ten companies with the highest market capitalization now come from the digital economy: Apple, Alphabet (Google), Microsoft, Amazon, Facebook and the Chinese corporations Tencent and Alibaba.[107] Of the 50 most-visited websites, the only non-commercial site is Wikipedia,[108] which consequently stands out as a lonely non-profit-making exception from the commercial mainstream.

These global corporations each operate in different sectors and each have monopolies in individual markets.[109] In 2019, more than 92 per cent of search queries worldwide were passed through Google[110] and its smartphone operating system, Android, has a market share as high as 76 per cent.[111] Facebook has similar market power: it was the first social network to surpass one billion registered accounts and currently sits at almost 2.38 billion monthly active users.[112] Facebook accounted for around 54 per cent of all social media site visits in the USA in August 2019 and holds 68 per cent of the social media market share worldwide.[113] It is not surprising that, in 2019, over 60 per cent of all US advertising revenue on the internet was divided up between these two corporations.[114] In China, Tencent dominates the social media with a market share of 79 per cent.[115] Amazon is the clear market leader in online shopping – with 47 per cent market share in the USA.[116] Alibaba, which could be called the Amazon of China, has a far tighter grip than Amazon's grip on the USA. According to the data, Alibaba will capture a 55.9 per cent share of all online retail sales in China this year.[117] Microsoft holds the monopoly in operating

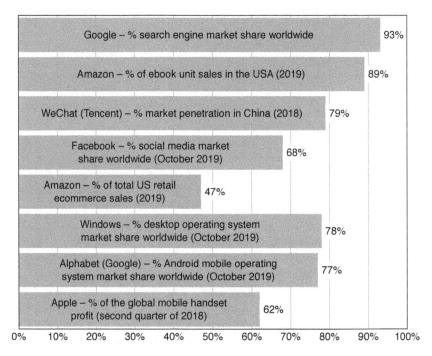

The major Internet corporations possess monopolistic market power in their respective lines of business. Apple pockets the lion's share of profits in the smartphone market, Alphabet dominates the operating systems for smartphones and Microsoft those for desktop computers. Amazon is especially powerful in online book sales and is on its way to dominate ecommerce sales, while Facebook and Tencent dominate social media.

FIGURE 4.4 Market power of IT companies

Sources: Counterpoint Research (2018); StatCounter (2019); eMarketer Editors (2019); Mansor Iqbal (2019); Day, M. and Gu, J. (2019).

systems on desktop computers with a 78 per cent market share.[118] And Apple has succeeded in dominating the smartphone market with its iPhones: the company generates 62 per cent of the global profits in this segment (see Figure 4.4).

How can we explain the fact that market power is so heavily concentrated around a few suppliers? Network effects and economies of scale are key reasons for their success: the more search queries Google processes, the better it can make its search results. And the more shopping people do on Amazon, the more diverse the range of products and the keener the prices that Amazon can offer. The result is the emergence of digital platforms for multiple services, which have a tendency to monopolize markets. Admittedly, they bring a great many suppliers together with a large number of users, but the coordination of this operates via one sole platform such as Netflix, eBay, YouTube, Instagram and others. In an utterly centralistic manner, they aggregate most of the internet traffic in their "markets" and rarely need to fear any serious competition.

Why is it that these platforms are operated by commercial suppliers rather than cooperative or even not-for-profit providers? One reason for this is the necessary investment required to build up a leading platform. Either companies are large enough to generate it themselves, or start-ups attract the necessary risk capital from venture capitalists. Since these investors demand a share in the company, and in future profits, this form of financing is unsuitable for cooperative or not-for-profit enterprises – in fact, it very much contradicts every basic principle of public-benefit oriented approaches.

An added factor is that the digital monopolists are now in a position to buy up potential competitors; some of them are enjoying phenomenally high profits. To give just a few illustrative examples: in the first quarter of 2018 alone, Alphabet had a net income of around US$31 billion; Microsoft's yearly net income in 2018/2019 was US$39 billion; Facebook made US$17 billion and the largest cloud platform, Amazon Web Services, made US$12 billion net income in the same period.[119] So, none of them are short of the "loose change" to fund a lavish shopping spree. For example, Alphabet bought YouTube for US$1.6 billion in 2006, Motorola's mobile communications division for US$12.5 billion in 2012 and the company Nest Labs for US$3.2 billion in 2014. Facebook acquired Instagram for roughly US$1 billion in 2012, Oculus for US$2 billion in 2014 and, in the same year, WhatsApp for a stunning US$19 billion. It is worth noting, incidentally, that these acquisitions are not driven purely by the giants' hunger for power. Being bought up is intrinsic to the professional ethos of the start-up scene. Dazzled by dollar signs, many start-up founders are not in the business of democratizing markets, nor indeed realigning them with ethical or environmental criteria by challenging established companies for their conventional business. Instead, the sale of their own start-up – ideally for hundreds of millions or preferably a sum in the billions – is often the real, albeit dismal, purpose of the business.

These buy-ups enable the giants to multiply their market power, partly because they can dominate several business segments. Initially, they focus on the market of their core business, as when one social network buys up another. At the same time, this allows them to enlarge their influence in the market for the analysis and evaluation of user data (big data), including consultancy services to third parties resulting from their data mining. And, finally, thanks to the enormous volumes of data to which they have access, the big operators gain advantages when it comes to establishing new business segments based on big data, such as the programming of learning algorithms and artificial intelligence.

Power in the digital economy is based, not only on large shares of particular markets for goods or service markets, but also on control of online infrastructures such as digital distribution channels. Amazon may not (yet) be a monopolist in the electronics, food or video markets; but, because so many users have stopped using search engines to search for products and now do so directly on Amazon, smaller providers are more or less forced to offer their books, videos and other products on Amazon's platform – and thus become dependent on the corporation's infrastructures.[120] Amazon gains the added advantage of comprehensive information on its competitors' sales trends and likely profits. A study discovered that, within the course of a few weeks, Amazon took many of the best-selling products from third-party providers into its own product range, a highly effective way of suppressing its competitors.[121] Nor is cartel law always followed to the letter. For instance, the EU Commission imposed a fine of €110 million on Facebook. When it purchased WhatsApp, the company had given assurances that it would not link user data across the two services. But, since 2016, Facebook has been doing exactly that. In the process, it considerably enlarged its monopoly-like position in the gathering and evaluation of personal information.[122]

A frequently cited counterargument to the monopolization thesis is that new companies could come along at any moment and sweep the old ones off the market with disruptive innovations. There is still every chance that new digital players will snatch away business from the established market giants, particularly in sectors where digitalization has not yet made major inroads, such as the food industry. But, it must not be forgotten that the only start-ups with real prospects of success are those that have the backing of well-financed venture capitalists. Many a struggle for market power is thus, not a David versus Goliath contest, but a battle between giants. And if one monopoly is simply replaced by another, that does very little for the democratization of the economy; and less still if the new internet giants concentrate even more power than the influential corporations of the past.

Perhaps the most important difference between monopolies in the analogue versus those in the digital economy is that the latter case always involves enormous power over the personal information of millions, if not billions, of people. The internet giants unify what used to be more stringently separated: financial power and increasing influence on information, news and public discourses. Therefore, monopoly building carries, not only economic, but also social, risks. Even now, the control wielded by monopolies over data and information is translating into political power – only a few years after many of these companies were founded. It is foreseeable that their economic, informational and political – and hence social – power will continue to grow. The consequences are almost impossible to overestimate. In Mach 2019, the video of the Christchurch attack was uploaded to Facebook via a mobile app, which was designed to allow extreme sports enthusiasts to stream footage from personal body cameras.[123] Copies of the Christchurch footage were downloaded and then circulated on other social media sites and found their way onto the front pages of some of the world's biggest news websites in the form of still images, gifs or even the full video.[124] The series of events around the Christchurch attacks has, once again, shone a spotlight on how social media sites try – and fail – to address far-right extremism on their platforms. Another critical view is taken of suspected attempts to influence the 2016 US presidential elections, not only by hackers and the Russian secret service, but also by social media. Even if Twitter, Facebook and others do not seem intent on pursuing any political agenda for the time being, the way in which they organize "filter bubbles" and echo chambers in social media is never entirely neutral.[125] And all of us have long been affected by the power they hold over our information. The German Spy Museum in Berlin has good reason to ask: "Who knows more about you – the Stasi, the NSA, Facebook or Miles and More?"[126]

At present, then, digitalization is not leading to economic democratization. On the contrary, a much more overwhelming trend is that IT and internet monopolies take advantage of network effects and huge financial volumes. The chances of economic democratization through decentralized structures, regionalization or cooperative organization seem to be slipping away. Alternatives still exist, though, and new platform cooperatives are emerging all the time with fantastic, forward-looking ideas for the transformation of our economy (see also Chapter 6). But, so far, these have remained within social niches.

The problems that massive concentrations of market power can unleash in relation to environmental issues and questions of justice are well-known from other markets, such as the chemical, pharmaceutical or food industries. Often representatives of the relevant corporations have direct roles within government, or government representatives take up jobs or consultancies in industry soon after leaving public office. This explains why many legislative attempts to introduce measures like higher environmental standards in these markets have failed or been heavily diluted (e.g. emission standards for the automobile industry in the EU). It, therefore, comes as no surprise that, particularly, the economic sectors where most power is concentrated – chemicals, agriculture, energy, and industry – have staved off the best efforts of sustainability-minded policy-makers and civil society for decades. If digitalization is to be shaped by democratic and social-ecological criteria, monopoly-law regulations must be passed as a matter of urgency, before the IT giants become too powerful in regulating policy.[135]

Technological utopia: blockchain

The World Wide Web revolutionized information exchange; then "Web 2.0" brought social media and the sharing economy. Now, blockchain technology could become an important building block of "Web 3.0" and decentralize and democratize the economy in a revolutionary way – or so the fans of this technology think.[127] It offers a tamper-proof way of exchanging money, goods and services online directly from person to person. It is mostly associated with cryptocurrencies such as Bitcoin, but it can also be used as an alternative for contracts, certificates and much more. In a nutshell, blockchain is a tool for simplifying transactions. Ultimately, it has the potential to eliminate the need for intermediaries such as central banking authorities, land registries and platforms like Airbnb or Uber. This is where its potential for democratization lies. Blockchain gives new impetus to libertarian dreams, ranging from an independent payment system via an uncensored internet through to complete anonymity on the web.[128]

It is not just peer-to-peer transactions that blockchain technology makes quicker and safer, but commercial trade as well. Apart from some initiatives from the alternative scene,[129] large companies are now shaping the development and application of blockchain. All ten of the largest public companies in the world, corporations as diverse as the Industrial and Commercial Bank of China, investment giant Berkshire Hathaway and computer giant Apple are exploring blockchain. A closer look reveals that at least 50 of the biggest names on the list have made their own mark on technology inspired by bitcoin.[130] So, is blockchain transforming capitalism or feeding into its hyper-acceleration? One way or another, there is a problematic conflict of objectives between democracy and environmental sustainability: blockchain's computational requirements are extremely energy intensive.[131] To calculate a single bitcoin, it takes about as much electricity as an average US household consumes in 21 days. The carbon footprint of a single bitcoin mined is equivalent to the carbon footprint of 743,677 VISA transactions or 49,578 hours of watching YouTube.[132] The electricity consumed by all Bitcoin transactions in 2016 may have been "only" 51 terawatt hours, according to a conservative estimate, but perhaps as much as 73 terawatt hours.[133] If Bitcoin were a country, it would rank 42nd in global power consumption – ahead of Nigeria, with a population of 180 million.[134] Although there are other blockchain applications that are less energy intensive, today entire data centres are commandeered for the technology's computations. Therefore, the dream of a decentralized economy based largely on blockchains would probably be a nightmare for the planet.

Polarization 4.0: less equality, less growth

In the preceding sections, we showed that digitalization within the current economic policy environment is breeding polarization in the labour market. We
also argued that digitalization of the economy carries a high risk of concentrating
market power and financial capital in the hands of just a few companies. Both
trends exacerbate economic inequalities, as we will explain in this section. For
wage inequality alone do not determine how income in a country is distributed.
Another important determinant is the breakdown between wage and capital
income. Here, too, digitalization comes into play.

   Assuming that there is no change in the regulatory framework of our economy, digitalization will be accompanied by a change in the ratio of wage to capital
income – a shift in favour of dividends, rental income, interest and share profits,
to be precise. How does that come about? The following train of thought explains
the macroeconomic mechanism of action. As we showed above, in the course of
digitalization, machines and algorithms replace human labour power. The more
machines and algorithms a firm deploys, the more revenues from production flow
to the owners of the firm.[136] It follows from all this that, because workers are being
replaced, wage incomes are falling; and, because more technologies are being used,
investment incomes are rising. Now that is not a hard-and-fast rule – the return on
capital (best visualized as the firm owner's income per robot) could just as well fall
and employees' hourly wages could rise. Because of the prevailing power relations
between business owners and employees, however, this is not what is happening
right now.[137] The example of self-driving cars illustrates these relationships: automated vehicles do not need drivers, but they do need hardware and software as well
as new infrastructure like server parks. When transportation, logistics or taxi companies make their drivers redundant, part of the money saved flows to the owners
of the self-driving car fleet and part to companies like Google, Tesla, Toyota or
Mercedes, and thus indirectly to those companies' employees. In these big businesses, however, the wage share is significantly lower than in small companies.[138]

   The effects of a full-scale digitalization of industry look no different: companies purchase robots because they can make clothing, furniture, cars or refrigerators more cheaply than their human forerunners. The cost-savings may be
passed on in the form of lower prices for these goods. But, another share lands in
the accounts of the owners of the "smart factories" and their (digital) technologies. Any number of examples can be listed: if newspaper articles are written by
algorithms, will the owners of the algorithms and the news platforms earn the
money, rather than the journalists? If financial advice is automated and there is far
less need for bank staff, the revenues still go to the financial institutions, but their
expenditure on employees etc. is lower. In short, digitalization turns out to be a
smart machine for bottom-to-top redistribution.

   For many decades, economists viewed the constant ratio between wage and
capital incomes as one of the stable elements of modern economies.[139] Since the

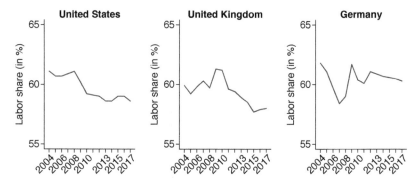

This graph shows the trends in the labour share in the USA, the United Kingdom and Germany from 2004 to 2017. The labour share is the share of total income accounted for by wages. The other part of total income is capital income, such as interest, dividends, share profits etc.

FIGURE 4.5   Trends in the wage share

Source: International Labour Organization (2019).

late 1980s, however, the share of income attributed to wages, known as the wage share, has been falling. More recent studies show that it has dropped significantly in the last few decades[140] – both in high-income and in low-income countries.[141] Figure 4.5 shows the recent trends in the USA, UK and Germany. Effectively, digitalization amplifies the proverbial Matthew effect: "to those who have, more will be given".

The shift from wage to capital incomes certainly cannot be explained solely by technological developments like digitalization. Since the dawn of industrialization, technical evolution has always involved the replacement of human labour power with machines and natural resources. But, on no account did that always lead to a decline in the wage share. For example, in the 1960s and 1970s, a period when industrial production was rapidly being automated, the wage share remained relatively constant and even rose slightly in Australia, Canada, Germany, France, Italy, Japan, Spain, the UK and the USA.[142] Experts, therefore, point to a second important correlation that explains shifts in the ratio of wage to capital income: the power relations between labour and capital – in other words, between employees and business owners or shareholders. The important maxim here is this: the greater the degree of employee organization, in trade unions for example, the higher the share of added value they can bargain for from the capital owners.[143] Indeed, in the 1960s and 1970s, the development of wages kept pace with rising productivity, partly because the trade unions in early industrialized countries were well organized and their negotiating position safeguarded by legislation. Added to that, companies' options to relocate production plants to other countries were more limited at that time: the balance of power also depends on the degree of globalization.[144]

This situation has changed drastically since the 1980s. Partly thanks to digitalization, businesses increasingly managed to shift production to countries of the global South where workers' wages are lower. At the same time, many countries' labour markets were being deregulated by a variety of political measures, another move that hampered trade unions from asserting their interests successfully. Real (or threatened) offshoring of production, as well as a declining degree of organization among employees, plays a part in the lower wage share even now. Today's crowd workers are barely unionized either,[145] which makes it easy for companies to play off the many homeworkers and self-employed workers against each other and keep wages down.

What is the impact of the increasing polarization of incomes on the dynamics of the entire economy; that is to say, on economic growth? This question can only be answered with reference to the trends mentioned above, particularly labour productivity. In Chapter 3, we showed how Industry 4.0 may boost labour productivity and, in many cases, also energy and resource efficiency and therefore promises high rates of growth. In sectors other than industry, the rationalization of workers can, in principle, be harnessed to extend product or service offerings. All the lorry drivers, industrial workers and office administrators who are replaced by robots and algorithms could either be redeployed to make even more freight deliveries, cars or services available, or they could produce different things. But this would imply high growth, which would be disastrous from an environmental perspective because higher production means higher demand for energy and resources. Rebound effects would thus eat up the potential savings from the efficiency gains.[146]

Clearly, the polarization of incomes and the precarization of low-income sections of the population are key reasons why we are most unlikely to return to high growth rates in the future. It could be said that digitalization is sawing off the branch on which it is sitting. For, while boosting resource efficiency and work productivity is one side of the digitalization coin, growing income inequality is the other. The latter will cause a gradual depression in consumption. If industrial workers lose their jobs, either they must live on transfer incomes, which are far lower, or else they must find new jobs, which tend to be less well paid. In both cases, their level of consumption will drop.[147] In general terms, many people's purchasing power will decline or stagnate, which means lower consumption. At the same time, of course, there are people whose incomes will rise, but these will be fewer in number and preponderantly higher earners, who usually spend a smaller proportion of their income on consumption (choosing to save instead). Overall, greater income inequality therefore results in a comparatively lower level of consumption for society as a whole than would be the case if income distribution was more equal. Depressed consumption in turn leads to lower investment by companies – and ergo to low economic growth. By way of heightening inequality, digitalization may contribute to sustained low growth rates.[148]

Indeed, many highly regarded economists now share the prognosis of sustained low growth. One of the first to put forward this thesis was American politician and professor of economics Larry Summers, addressing the world's economic experts at the annual conference of the International Monetary Fund in November 2013. Summers declared that the low growth rates since the financial crisis of 2007/2008 were not a passing phase. Instead, we could expect a lasting phase of low growth, known as "secular stagnation".[149] A similar case has since been made by other prominent economists, such as Paul Krugman and Robert J. Gordon.[150] Their analyses coincide with empirical studies that find a sustained long-term decline in growth rates in high-income countries.[151]

Then again, is low growth not a good thing from an environmental viewpoint? Could it be that we have already entered the "post-growth economy",[152] which parts of civil society have long been demanding in order to solve the major social and environmental crises of our epoch? Far from it! The social-ecological and anti-growth models being discussed under the heading of post-growth or degrowth conjure up a distinctly different picture from the state of affairs we see at present.[153] Far from being democratic, egalitarian and environmentally sustainable, the current no-growth economy is characterized by rising inequalities, growing concentration of power and an ecological footprint far in excess of the sustainable level. Rather than a "post-growth economy", the situation could therefore more accurately be described as the beginnings of "neo-feudalism",[154] – i.e. an economic situation in which people have little or no chance of attaining prosperity through work and achievement, but almost exclusively through wealth-building and capital income. As we have shown, uncontrolled digitalization of the economy runs the risk of further amplifying the present trends towards income polarization, precarization and the concentration of wealth and economic power. Must we envisage the digital economy of the future as "digital neo-feudalism"?

Free riding on public goods

Democracy and participation in (economic) policy decision-making by their very nature countermand a neo-feudalist system because both concentration of power and ever-growing economic inequalities are incompatible with the requirement for an equitable voice and participation for all. We now turn to another aspect that feeds into the social and political imbalances described: many digital companies are profiting from the given infrastructures and common goods of a country or region without making an appropriate contribution to their upkeep. For instance, Airbnb earns income by acting as an agent for holiday apartments but does not pay any tourist tax. Uber dispatches drivers onto the roads in many countries, but contributes negligible amounts of tax towards road-building and maintenance.[155] And Google makes billions from advertising, but pays next to nothing to the exchequer. Do we have a digital free-rider problem here?

˙Technological utopia: floating cities

The idea of building floating cities to create a separate physical and legal space has been circulating in the IT industry for some time now. The activities of the Seasteading Institute, whose financial backers include Peter Thiel, a well-known investor from Silicon Valley, are a hot topic of discussion.[161] Since floating cities on the open sea would not be subject to any national laws, they could be a dream come true for libertarians.[162] A first prototype was due to be launched near French Polynesia in 2018.[163] Although not yet fully independent from government authority, it still enjoys many freedoms.[164]

Despite pockets of opinion that consider these islands eco-friendly[165] because their power consumption can be covered by renewable energy sources,[166] "seasteading" looks set to become an extremely resource-intensive project. Apart from being constructed from steel, concrete, glass and other materials, the artificial islands are laden with high-tech devices. And even that is not the most serious problem. Floating IT companies would be definitively free of any tax burden and the "old territorial states" would be left out in the cold.

In the previous sections, we showed why digital businesses are currently growing at such a pace and can easily turn into quasi-monopolistic, globally operating corporations. It is no secret that global players are adept at using their power to keep wages low by pointing to international competition and hinting at the possibility of offshoring production. But, they are equally capable of circumventing taxes, as countless examples including Ikea,[156] Starbucks[157] and Apple[158] demonstrate. Of course, under capitalism, businesses try to keep their tax payments as low as possible in order to maximize their profits. But, the opportunities of digitalization are helping them to pursue this objective by far more perfidious methods. Many internet corporations play this game extremely successfully. In order to avoid taxes, complicated accounting models operating across several countries are chosen. Other internet companies, besides Google, also make use of the now familiar "double Irish" tax-saving model. The principle is not to account for profits in the countries where they are earned, but where the tax rates are especially low.[159] The European commission estimates that digital businesses paid an effective average tax rate of only 9.5 per cent, compared with 23.2 per cent for bricks-and-mortar firms in 2018.[160] But, why are internet firms everywhere allowed to use telephone, data cables and other infrastructures for free, without paying a fair share towards their upkeep – let alone contributing to other public institutions like schools, universities and hospitals, which are financed from tax revenues?

It tends to be even more difficult for the state to collect taxes from digital companies than from other corporations. A company like IKEA at least has to open retail branches, with doors where the tax authorities can come knocking. Exactly the same goes for the Starbucks chain: it needs real cafés to sell its coffee.

For businesses like Spotify or Bing in the digital economy, matters are more difficult. For, if there are no Spotify outlets and no staff in a country, how can the fiscal authorities present the business with a tax demand? In recent years, the EU has debated a "digital tax", addressing large international companies. As many of such companies are from the USA, it is not surprising that this proposal has led to a dispute with US President Donald Trump.[167] However, little results have been achieved yet. Only France has introduced such a digital tax – albeit at a very low rate of 3 per cent.[168] Implementing such a tax for the entire EU has been prevented by the opposition of several countries, such as Luxembourg, who feared a taxation of financial services, or Ireland, the home of Google's, Facebook's and Microsoft's European headquarters, or Sweden and Denmark. The German government is now pushing for minimal corporate taxes in the OECD instead of implementing a digital tax. While making sense in principal, the level of taxation is very low – and even this low level is uncertain to be actually put into place as there are many political challenges both within countries and among countries.[169]

Although, to some extent, the virtual economy can be everywhere and nowhere, it tends to concentrate in certain cities or regions.[170] Silicon Valley is the most famous example. There may be many reasons why the hotspot of digitalization is there in particular – one of them is certainly the relatively low taxation. This becomes even more obvious in the locations chosen by the internet giants for their European branch offices. Many have based these in Ireland, as already mentioned, where tax demands are especially low. In 2014, Apple reportedly paid taxes of only 0.005 per cent on its European profits, which the corporation declares solely in Ireland, home to its European headquarters. In the meantime, the EU Commission concluded that Apple had paid substantially less taxes in Ireland than it should have for years by clever financial engineering and has declared Apple's low tax rates an illegal subsidy under a state aid procedure, and is demanding supplementary payments. In 2016, the Commission ordered the US company to pay €13 billion (US$14.4 billion) of taxes to Ireland. Nevertheless, Ireland refuses to levy more taxes and is appealing against the decision at Europe's General Court.[171] Although the country has a corporate tax rate of 12.5 per cent (lower than America's 21 per cent), its generous corporate tax planning tools allow foreign companies to achieve an effective tax rate of between 0 and 2.5 per cent on global profits rerouted to Ireland though tax treaty networks.[172]

The IT giants have long ranked among the ten largest companies on the stock exchange, but barely contribute to the financing of public goods. This makes it ever more difficult for the state to redistribute from rich to poor. The declining redistribution effect of state taxation can also be observed empirically.[173] Be it in

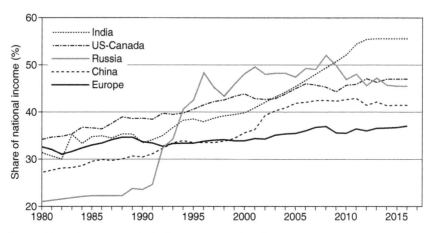

Income inequality rose markedly in many countries. The graph shows how the income shares of the top 10% have risen between 1980 and 2016. It increased significantly in all countries and country groups depicted here.

FIGURE 4.6 Trends in income inequality

Source: World Inequality Lab (2018).

the labour market, the distribution of national income or the payment of taxes, digitalization is currently helping to make society more unfair. Of course, this development has been under way for around three decades. If we compare the Gini coefficients – a common unit of measure for income inequality – of all OECD countries between 1985 and 2013, inequality is found to have risen in 17 out of 22 countries. In four countries it remained unchanged and in only one country, Turkey, did it decrease.[174]

To be sure, other forces are also driving the rise in inequality – above all, economic policy measures in the fields of fiscal and labour market policy and, of course, globalization. Digitalization is just one factor and perhaps not even the most important. But, its effect is still significant and the direction is clear: under present conditions, it contributes to less rather than more economic justice. So, if our society is going to become even more digital in future, a distinctly different policy environment must be created to make a fair digital society possible. In Chapter 6, we will look at such a "transformative digital policy". But, first, we address another question: even if digitalization does not make our society fairer – does it nevertheless enhance our freedom and happiness as human beings?

Living a speedy or "good life"?

Who can imagine life without the internet anymore? How would we plan our trips, for example? We would have to negotiate railway journeys without a travel app or road trips without a navigation system. How could we send someone a quick update if we were running late or needed to change the agreed meeting place at short notice? And, if we had to forego digital mapping services like OsmAND or Google Maps, can we even remember how to use analogue street maps as a quick and efficient method to find our way to an unknown address? It is not just the "digital natives" from the generations born after 1980[175] who find it hard to imagine how our working and everyday lives can be organized without computers. But, the question is, does digitalization – despite its tendency to increase economic injustices – nevertheless bestow so much freedom and autonomy that it makes us happier overall?

Countless examples suggest that it does. Most economists would intuitively agree with this assumption because they evaluate the prosperity of a country primarily by the services and consumption options that are available. For example, that is why in their book, "The Second Machine Age", Erik Brynjolfsson and Andrew McAfee use time travel as a thought experiment to illustrate how new products make us happier.[176] They ask what you would rather do: spend $50,000 on a basket of products from the year 1993, or on the equivalent products from 2013? And they are certain you would choose to be able to buy the 2013 products. There is a reason for this: a whole range of products exist today that were not available back then, but which we now consider indispensable for a good life. Nowadays, who would want to fast-forward on an audio cassette searching for the beginning of the next song rather than having one-click access to any song at all? Who still cares to watch puppet films when animated movies can be viewed in 3D? And which driver prefers to check the road atlas for the next exit instead of relaxing and letting automatic announcements guide them to their destination?

The thought experiment makes two key mistakes, however. First, we cannot turn the clock back. In the 1980s, mobile phones were still unknown, let alone smartphones, so it was no hardship to do without them. Second, and more importantly, it makes no sense to ask whether an individual would forego the new digital technologies, because structures and customs have shifted so much that it is virtually impossible to participate in social life without a computer or smartphone. Therefore, rather than asking individuals, society must be studied in its entirety: are societies with digital technologies happier today than they were in 1993 without them?

It is far from easy to determine the well-being of a society using scientific methods. Two approaches have taken shape. The first assumes that people themselves are very good at assessing whether they are happy or not. Over many years, repeated surveys asking questions like "How satisfied are you with your life overall?" have concluded that well-being in high-income countries has not risen for quite some time. The contentment of populations in these countries seems to have reached a zenith. And this was the level attained by the USA shortly after the Second World War[177] – long before digitalization found its way into our workplaces and our lives. In fact, the median per capita income is the highest in US history, the per capita purchasing power has almost doubled in the USA since 1967[178] and the same span of time saw many aspects of the workplace and everyday life turned upside down – first by personal computers, then the internet, by mobile phones at the turn of the millennium and smartphones since 2007. Yet, apart from minor fluctuations, the reported level of well-being has remained static during this period.

The second method for measuring the well-being of a society attempts to do so using "objective" factors. These are aspects that are assumed to exert a great influence on people's welfare. To this end, simple models make use of the disposable income or the gross domestic product of a country. But, since money alone is not an adequate indicator of prosperity, additional factors are frequently referred to. In one wide-ranging study, British academics Kate Pickett and Richard G. Wilkinson used life expectancy, educational level, infant mortality, prison population numbers, psychological illnesses and other social and health-related problems as indicators for social well-being. Their findings suggest that there is no positive correlation between the level of gross domestic product and welfare in prosperous countries.[179] At the same time, the study shows that higher levels of inequality correlate with poorer outcomes in relation to the social and health-related problems mentioned.

If digitalization continues to increase inequality, this will have negative impacts on welfare. Many signs of this are evident in the developments outlined so far. It is well known that, for people living in both relative and absolute poverty, the probability of being unsatisfied with life is higher, and – particularly for children – health-related problems increase. Meanwhile, the rise in precarious employment engenders a rising fear in the lower-middle class that they, too, could slide into poverty. And the phenomenon of "social acceleration" is considered to be especially relevant to individual well-being within the context of digitalization.[180] Digitalization undoubtedly supports swift and agile reactions in everyday situations, supplies information about faster journey times, makes it possible to shop around the clock and gives almost instant access to all sorts of information. But, if digital tools help us to save so much time, then why has society not registered a marked rise in well-being? Surely a marvellous new era of time prosperity should be dawning? One reason why this is not happening is a phenomenon known as time-rebound effects.[181] These are manifested in our tendency to fill time with new and additional activities, negating potential time-savings in the process. And, structurally, they stem from the ever-increasing pace of society in general.[182] So, let us take a closer look at the relationship between digitalization, the individual pace of life and social acceleration.

The acceleration effect of digitalization is most clearly evident in interpersonal communication. It has brought about a radical surge in the volume of messages we send. The more quickly and easily we can communicate, the more we do so. Over half of the world population used email in 2019. The total number of business and consumer emails sent and received per day will exceed 293 billion that

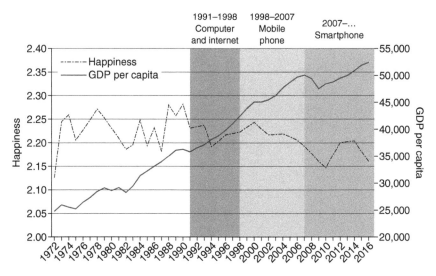

Despite economic growth and the diffusion of various information and communications technologies, there has been no increase in people's average happiness in the US. Income per capita has more than doubled since 1972 while happiness has remained roughly unchanged or has even declined. The diffusion of important digital technologies took place over the same period of time.

FIGURE 4.7 Economic growth, digitalization and satisfaction with life

Source: Sachs (2018).

year.[183] Every minute, users send over 18 million text messages, watch 4.5 million videos on YouTube, send over 500,000 Tweets on Twitter, post over 50,000 new photos on Instagram and swipe 1.4 million times on Tinder.[184] As of January 2019, the internet reaches over 56 per cent of the world's population – a 9 per cent increase from January 2018.[185] The explosive growth in the total volume of electronic messages we exchange has become impossible to quantify. The mere fact that we now write and receive many more messages every day is accelerating our lives, because, the more time we spend communicating, the less time we have for other tasks. And these do not decline in number, rather the opposite. Consequently, we have to cram more tasks into shorter time frames.

Then again, when it comes to dealing with this concentration of tasks, digital tools can be a tremendous help – in three ways.[186] First, we can replace time-consuming, "analogue" actions with time-saving, digital equivalents: writing an email instead of a hand-written letter, for instance, or buying online rather than driving to the shops. Second, digital communication accelerates the pace of activities by reducing the length of breaks and idle times. For instance, a several-day wait for a postal reply – on which a subsequent action depends – is shrunk down to a few seconds by electronic communication. And enforced breaks while travelling on public transport or waiting for the next appointment are all but a thing of the past. More and more often, smartphones are used to fill these chunks of time, sometimes productively and sometimes less so. Third, not only is the speed of communication and activities on the rise, but digital tools make multitasking even easier. Thanks to emails, texts, WhatsApp, feeds, chats, posts and so on, information generated at different times can be received and responded to instantly and translated into subsequent actions – even when the other person is far away, doing something completely different or even asleep. And, in the middle of meeting friends in a café, we can upload photos to Instagram or Snapchat and simultaneously "cultivate" other friendships on the other side of the globe. This is how the transition from the analogue to the digital society draws us into an escalating frequency of activities and experiences per unit of time.[187]

Digitalization is also driving extreme acceleration in economic affairs: applications such as barcoding, big-data analyses of consumer behaviour or just-in-time marketing can efficiently match product cycles with social trends and fashions. This shortens product life cycles and fashion cycles. Algorithmic stock-market trading organizes the exchange of shares, bonds and other financial products in nanoseconds, leading to more and quicker financial transactions and keeping the real economy on its toes. The high-speed economy pervades the whole of the value chain, right through to the individual workplace: was there an unanswered email in your inbox for two hours? If so, you will probably receive the first reminder soon, and then a concerned inquiry about whether you have changed jobs.

Little wonder that the main complementary phenomenon of acceleration is time scarcity.[188] The greater the acceleration of life and economic processes, the more apparent the downsides become: we do indeed reach our destination more quickly, but we spend less time there. We arrange more meetings with work colleagues or friends, yet spend part of the time on our mobile phones arranging the next meeting rather than talking to them. The attention we might pay to others is devoured by the attention we pay to our smartphones. And, altogether, restlessness abounds in daily life. If digital tools make it easier for us to meet with others, yet the time we spend together is shorter and fleeting in quality, then the rise in digital options does not boost personal well-being or even happiness. So, ultimately, the greatest threat posed by digitalization is its very success.

Even if hardly anyone would be willing (or able) to give up their smartphone, for some people being constantly available and bombarded with electronic messages feels like stress. And more stress arises when digitalization contributes to blurring any clear distinction between work and leisure time. It is not just a matter of how many emails are written at home in the evening or how long our time off after work was disrupted. The mere expectation of work-related emails after hours is enough to send the stress level soaring.[189]

It is not even unusual for stress to turn into addiction. Online addiction manifests itself in different ways, such as internet addiction (surfing, unfocused clicking), mobile phone addiction (texting, constantly checking for calls) or social media addiction (chat addiction), and these can be overlapping phenomena.[190] According to an analysis of 31 nations across seven world regions, 2 to 11 per cent of the population are considered to be online addicts and young people are particularly affected.[191] The signs of addiction for those affected are the feeling of a constant urge to be online or check their smartphone, the fear of missing out and a sense of emptiness and frustration if they have been offline for any length of time. This leads to diminished attention spans at school or at work. A vicious cycle can begin if declining concentration and dissipation of effort cause a build-up of performance pressure, triggering new stress.

So, what can be said, on balance, about digitalization, acceleration and well-being? Is there a parallel here with the effects of digitalization on the environment? In Chapter 3, we showed that the potential of digitalization to save energy and resources is offset by the large ecological footprints of digital devices and infrastructures, and because it leads to rebound effects and economic growth. We, therefore, concluded that the environmental effects amount to a zero-sum game at best. Could the same conclusion be drawn about personal well-being in the digital age? The happiness promised by the increasing options and improved convenience of digitalization and the many ways it makes everyday life and communication easier is offset by the accelerating pace of life and rising performance pressure and stress levels – even to the point of burnout in growing numbers of cases. And the treadmill never stops: many people feel under pressure to max out on their digital options with ever more communication and ever more activities. The net result is a very real risk that, unless we are careful and take mitigating action, the potential for a better quality of life will be cancelled out.

Considering these ambivalent effects of digitalization on the quality of life, in no way can they be said to counterbalance the negative consequences for economic justice. The effects of polarization in the labour market and in income, combined with the concentration of economic and social power, result in a society that is considerably more unjust. All in all, the digitalization that is happening today is clearly not making the majority of people better off. Yet, how digitalization plays out is not God-given, but hinges on how we as a society choose to shape it. We firmly believe that digitalization can be different! In the next two chapters we outline the shape that it might take.

Notes

1 Hobsbawm, The machine breakers, 1952.
2 Schumpeter, *Capitalism, socialism and democracy*, 1942.
3 Bartmann, *The Return of the Servant*, 2016.
4 Muntaner, Digital Platforms, Gig Economy, Precarious Employment, and the Invisible Hand of Social Class, 2018.
5 Schwarz, How Many Photos Have Been Taken Ever?, 2012.
6 Reuters, Instagram's user base grows to more than 500 million, 2016.
7 *The Economist*, The last Kodak moment?, 2012.

8 While the 145,000 jobs relate to the year 1988, the peak sales figure was achieved in 1996. *The Economist*, The last Kodak moment?, 2012.
9 Apple, Annual Report, 2016.
10 Aslam, Instagram by the Numbers, 2017.
11 Lucas, *The search for survival*, 2012.
12 Klemm, Stressiger Beruf, 2016.
13 Hillebrand *et al.*, Technology and change in postal services – impacts on consumers, 2016.
14 Bureau of Labour Statistics, Occupational Employment Statistics, 2018.
15 Gershgorn, DRIVING AWAY, 2016.
16 Levy, Can an Algorithm Write a Better News Story Than a Human Reporter?, 2012.
17 Muro/Maxim/Whiton, Automation and Artificial Intelligence, 2019.
18 See e.g. Ginder *et al.*, Enrollment and Employees in Postsecondary Institutions, 2017; Lederman, Who Is Studying Online (and Where), 2018.
19 Frey/Osborne, The future of employment, 2013. The authors do not make any precise statements about how rapidly this process could unfold, speaking instead relatively vaguely of one to two decades.
20 Zierahn/Gregory/Arntz, The risk of automation for jobs in OECD countries, 2016b.
21 International Federation of Robotics, The Impact of Robots on Productivity, Employment and Jobs, 2017.
22 Zierahn/Gregory/Arntz, Racing With or Against the Machine?, 2016a.
23 World Economic Forum, Future of Jobs Report, 2018.
24 Acemoglu/Restrepo, *Robots and Jobs*, 2017; further studies include Acemoglu/Restrepo, *The race between machine and man*, 2016; Brynjolfsson/McAfee, The Second Machine Age, 2014.
25 Muro/Maxim/Whiton, Automation and Artificial Intelligence, 2019.
26 Charniak/McDermott, Introduction to Artificial Intelligence, 1985.
27 Sullivan, How Machine Learning Works, As Explained By Google, 2015.
28 Bögeholz, Künstliche Intelligenz, 2017; Silver *et al.*, Mastering the game of Go without human knowledge, 2017.
29 Purdy/Daugherty, Why Artificial Intelligence is the future of growth, 2016.
30 Brynjolfsson/McAfee, The Second Machine Age, 2014.
31 Strubell, Energy and Policy Considerations for Deep Learning in NLP, 2019.
32 World Economic Forum, With lifelong learning, you too can join the digital workplace, 2019.
33 World Economic Forum, 5 things to know about the future of jobs, 2018a.
34 Oxford Economics, How robots change the world, 2019.
35 PricewaterhouseCoopers, Will robots really steal our jobs, 2018.
36 PricewaterhouseCoopers, Sizing the prize, 2017.
37 World Economic Forum, Future of Jobs Report, 2018.
38 Brynjolfsson/McAfee, Race Against the Machine, 2012.
39 Vardi, Moshe, Humans, Machines, and Work, 2017.
40 Feenstra/Inklaar/Timmer, The next generation of the Penn World Table, 2015; The Economist Data Team, "Secular stagnation" in graphics, 2014.
41 World Economic Forum, Future of Jobs Report, 2018.
42 Norton, Automation, the changing world of work, and sustainable development, 2017a.
43 Norton, Automation and inequality, 2017b.
44 Dolan/Peasgood/White, Do we really know what makes us happy?, 2008.
45 Petschow *et al.*, Social well-being within planetary boundaries, 2018.
46 Brynjolfsson/McAfee, The Second Machine Age, 2014.
47 Witte, The effects of automation on unemployment and mental health, 2019
48 Merchant, Fully automated luxury communism, 2015.

49 For a typology of Crowdwork Platforms see: Howcroft/Bergvall-Kåreborn, A Typology of Crowdwork Platforms, 2019.
50 Huws/Spencer/Joyce, Crowd Work in Europe, 2016.
51 Huws/Spencer/Joyce, Crowd Work in Europe, 2016.
52 For more examples see: https://requester.mturk.com/create/projects/new.
53 Marx, Die deutsche Ideologie, 1969, p. 33 (quoted in English after Karl Marx. The German Ideology. www.marxists.org/archive/marx/works/1845/german-ideology/ch01a.htm, accessed: 10 November 2018).
54 Schor/Attwood-Charles, The sharing economy, (2017).
55 Cantarella/Strozzi, Labour market effects of crowdwork in US and EU, 2018.
56 Berg, Income security in the on-demand economy, 2015.
57 Huws/Spencer/Joyce, Crowd Work in Europe, 2016.
58 For examples of unions from the UK, Germany, Austria, Sweden and the USA defending the rights of crowd- and platform-workers see: http://faircrowd.work/unions-for-crowdworkers/ (accessed: 22 October 2019).
59 Cantarella/Strozzi, Labour market effects of crowdwork in US and EU, 2018.
60 Adams/Berg, When home affects pay, 2017.
61 Adams/Berg, When home affects pay, 2017.
62 Overall there is still little research on the relationship between gender relations and digitalization.
63 See e.g. China Labor Watch, Apple making big profits but Chinese workers' wage on the slide, 2016; Sacom, Workers as Machines, 2010.
64 Brand/Wissen, *The limits to capitalist nature*, 2018; Kopp *et al.*, *At the Expense of Others?*, 2019.
65 Walker, A Map of Every Device in the World That's Connected to the Internet, 2014.
66 Huws/Spencer/Joyce, Crowd Work in Europe, 2016.
67 See e.g. Calo/Rosenblat, The taking economy, 2017; Lee *et al.*, Working with machines, 2015.
68 Howcroft/Bergvall-Kåreborn, A Typology of Crowdwork Platforms, 2019.
69 Leimeister/Durward/Zogaj, Crowd Worker in Deutschland, 2016.
70 For an example on how companies and organizations are working on creating an "assembly line for knowledgework" see e.g. Ipeirotis/Little/Malone, Composing and analysing crowdsourcing workflows, 2014.
71 The World Development Report (WDR), The Changing Nature of Work studies, 2019.
72 World Economic Forum, Future of Jobs Report, 2018.
73 Bartmann, The Return of the Servant, 2016.
74 Bartmann, The Return of the Servant, 2016.
75 For the listed data see: Bureau of Labour Statistics, Contingent and Alternative Employment Arrangements News Release, 2018b. For debates about the rising precariat see e.g.: Greenstein, The Precariat Class Structure and Income Inequality Among US Workers, 2019; Standing, The Precariat, 2018.
76 Ahlers *et al.*, Genderaspekte der Digitalisierung der Arbeitswelt. Diskussionspapier für die Kommission "Arbeit der Zukunft", 2017.
77 Turner, *From counterculture to cyberculture*, 2010.
78 Ottmann, Liberale, Liberal, Republican and Deliberative Democracy, 2006.
79 Fawkes/Gregory, Applying communication theories to the Internet, 2000.
80 Pariser, *The filter bubble*, 2011.
81 Habermas, *Ach, Europa*, 2014, pp. 161f., own translation.
82 Welzer, Smart Dictatorship, 2016.
83 Lanier, Ten arguments for deleting your social media accounts right now, 2018, p. 22.
84 Lanier, Ten arguments for deleting your social media accounts right now, 2018, p. 5.
85 Paris Innovation Review, Agriculture and food, 2016.
86 For more examples see e.g. Paris Innovation Review, Agriculture and food, 2016.

87 Scholz, Platform Cooperativism, 2016a.
88 Retrieved from http://backfeed.cc/ (accessed: 19 November 2019).
89 Stallman, *The GNU manifesto*, 1985.
90 Gates, An open letter to hobbyists, 1976.
91 Skerrett, IoT Developer Survey 2017, 2017, p. 23.
92 StatCounter, Global market share held by leading desktop internet browsers from January 2015 to June 2019, 2019b.
93 Schumacher, Small is beautiful, 1973.
94 Hines, *Localization*, 2000; McKibben, *Deep economy*, 2007; Sachs/Santarius, *Slow Trade*, 2007.
95 Lange, Beyond A-Growth: Sustainable Zero Growth, 2019; Latouche, *Farewell to growth*, 2009; Sekulova/Schneider, Open Localism, 2014.
96 Sachs/Santarius, *Slow Trade*, 2007.
97 Wikipedia.org, Distributed economy, 2017b.
98 See e.g. Altieri, *Agroecology: The science of sustainable agriculture*, 1995.
99 Keppner *et al.*, Focus on the future: 3D printing Trend report for assessing the environmental impacts, 2018.
100 See e.g. Rösch/Dusseldorp/Meyer, Precision Agriculture, 2005; Nova-Institute for Ecology and Innovation, High-tech strategies for small farmers and organic farming, 2018.
101 Gershenfeld, How to make almost anything, 2012.
102 Petschow *et al.*, *Dezentrale Produktion, 3D-Druck und Nachhaltigkeit,* 2014.
103 Ferdinand *et al.*, *The Decentralized and Networked Future of Value Creation*, 2016.
104 Csikszentmihályi, Making A Fresh Start, 2017.
105 Liu *et al.*, Sustainability of 3D Printing, 2016.
106 Kreiger *et al.*, Life cycle analysis of distributed recycling of post-consumer high density polyethylene for 3-D printing filament, 2014; Rifkin, *The zero marginal cost society*, 2014.
107 Staley, There's a new list of the world's 10 largest companies—and tech isn't on it, 2018.
108 Wikipedia.org, Wikipedia, 2017a.
109 Canaccord Genuity, Apple Claims 92% of Global Smartphone Profits, no year; Bundesverband der deutschen Versandbuchhändler, *Statista*, 2015; The Economist, Android attack, 2016.
110 StatCounter, Search Engine Market Share Worldwide, 2019d.
111 StatCounter, Mobile Operating System Market Share Worldwide, 2019c.
112 Statista, Most popular social networks worldwide as of October 2019, ranked by number of active users, 2019b.
113 StatCounter, Social Media Stats Worldwide, 2019; Statista, Most popular social network websites in the United States in August 2019, based on share of visits, 2019.
114 Statista, Amazon Challenges Ad Duopoly, 2019a.
115 Mansor Iqbal, WeChat Revenue and Usage Statistics, 2019.
116 eMarketer Editors, Digital Investments Pay Off for Walmart in Ecommerce Race, 2019.
117 Gupta, Just how far ahead is Alibaba in China's e-Commerce market, 2019.
118 StatCounter, Desktop Operating System Market Share Worldwide, 2019a.
119 Data retrieved from www.macrotrends.net/ on 21 November 2019.
120 Mitchell, Amazon Is Trying to Control the Underlying Infrastructure of Our Economy, 2017.
121 Feiner, Amazon admits to Congress that it uses 'aggregated' data from third-party sellers to come up with its own products, 2019.
122 Reuters, EU fines Facebook 110 million euros over WhatsApp deal, 2019.
123 Hern/Waterson, Social media firms fight to delete Christchurch shooting footage, 2019.

124 Wakefield, Christchurch shootings, 2019.
125 Pariser, *The filter bubble*, 2011; Kelly/Françoise, This is what filter bubbles actually look like, 2018.
126 Deutsches Spionagemuseum Berlin, Deutsches Spionagemuseum Berlin – German Spy Museum Berlin, 2019.
127 Zago, Why the Web 3.0 Matters and you should know about it. Medium, 2018.
128 Bollier, Will Bitcoin and Other Insurgent Currencies Reinvent Commerce?, 2014.
129 See e.g. Ethereum, Ethereum. Blockchain App Platform, 2017; for an overview of other blockchain players, see Patrizio, 35 Blockchain Startups to Watch – Datamation, 2017.
130 del Castillo, Big Blockchain, 2018; for an overview of other blockchain players, see Patrizio, 35 Blockchain Startups to Watch – Datamation, 2017.
131 Greenfield, Radical technologies, 2017.
132 Digiconomics, Bitcoin Energy Consumption, 2019.
133 Digiconomics, Bitcoin Energy Consumption, 2019.
134 The country closest to Bitcoin in terms of electricity consumption is Austria.
135 Zuboff, *The age of surveillance capitalism*, 2019.
136 Brynjolfsson/McAfee, Race Against the Machine, 2012.
137 Piketty, *Capital in the Twenty-First Century*, 2014. See also the discussion of altered power relations further below.
138 Autor *et al.*, Concentrating on the Fall of the Labor Share, 2017.
139 See for an early discussion Kaldor, A model of economic growth, 1957.
140 International Labour Organization, *Wages and equitable growth*, 2013.
141 Trapp, Measuring the labour income share of developing countries, 2015.
142 International Labour Organization, The LabourShare in G20Economies, 2015.
143 Bentolila/Saint-Paul, Explaining movements in the labor share, 2003.
144 Stockhammer, Determinants of the Wage Share, 2017.
145 Scholz, Uberworked and underpaid, 2016b.
146 Lange *et al.*, *Economy-Wide Rebound Effects*, 2019.
147 Cf. Staab, The consumption dilemma of digital capitalism, 2017.
148 Staab, The consumption dilemma of digital capitalism, 2017.
149 Summers, Larry Summers at IMF Economic Forum, 2013.
150 Teulings/Baldwin, *Secular Stagnation*, 2014.
151 Lange/Pütz/Kopp, Do Mature Economies Grow Exponentially?, 2018; Wibe/Carlén, Is Post-War Economic Growth Exponential?, 2006.
152 Lange/Jackson, Speed up the research and realization of growth independence, 2019.
153 D'Alisa/Demaria/Kallis, *Degrowth: a vocabulary for a new era*, 2014; Jackson, *Prosperity without Growth*, 2016; Lange, *Macroeconomics Without Growth*, 2018.
154 See e.g. Hudson, The Road to Debt Deflation, Debt Peonage, and Neofeudalism, 2017; Morozov, Tech titans are busy privatising our data, 2016; Pawel, You Call It the Gig Economy. 2019.
155 This definitely applies to Belgium, see Charlot, Uberize me, 2016.
156 Auerbach, Ikea: Flat Pack Tax Avoidance, 2016.
157 Campbell/Helleloid, Starbucks: Social responsibility and tax avoidance, 2016.
158 Duhigg/Kocieniewski, Apple's Tax Strategy Aims at Low-Tax States and Nations, 2012.
159 See on this the *New York Times*, 'Double Irish With a Dutch Sandwich', 2012.
160 Rankin, Facebook, Google and Amazon could pay 'fair' tax under EU plans, 2018.
161 Wong, Seasteading, 2017.
162 Steinberg *et al.*, Atlas Swam, 2012.
163 Carli, Oceantop Living in a Seastead, 2016.
164 Wachs, Seasteaders to bring a libertarian floating community to the South Pacific, 2017.
165 Carli, Oceantop Living in a Seastead, 2016.

166 Cf. Carli, Oceantop Living in a Seastead, 2016.
167 Greive/Hildebrand/Berschens, Trump set to tussle over EU digital tax plans, 2018.
168 Hermant, INSIGHT: France Taxes the Digital Economy, 2019.
169 Lobb/Stack, G20/OECD work: Maybe way beyond digital, 2019.
170 Shearer/Vey/Kim, Where jobs are concentrating and why it matters to cities and regions, 2019.
171 Foo Yun Chee, Apple spars with EU as $14 billion Irish tax dispute drags on, 2019.
172 Cao, Apple Refuses to Pay Ireland $14 Billion in Back Taxes, 2019.
173 Piketty/Saez, How progressive is the US federal tax system?, 2007.
174 Keeley, *Income Inequality*, 2015.
175 Palfrey/Gasser, *Born Digital*, 2008.
176 Brynjolfsson/McAfee, *The Second Machine Age*, 2014.
177 Easterlin, Does economic growth improve the human lot?, 1974; Easterlin/McVey, The happiness-income paradox revisited, 2010.
178 Amadeo, Income Per Capita, With Calculations, Statistics, and Trends, 2019.
179 Wilkinson/Pickett, *The Spirit Level*, 2010.
180 See in general on this e.g. Druckman *et al.*, Time, gender and carbon, 2012; Gleick, *Faster. The Acceleration of Just About Everything*, 1999; Rosa, *Social acceleration*, 2013; Rosa, Social acceleration, 2003; Santarius, Energy Efficiency and Social Acceleration, 2016b.
181 Binswanger, Technological progress and sustainable development, 2001.
182 See in general on this e.g. Linder, The harried leisure class, 1970; Rosa, *Social acceleration*, 2013; Schulze, The experience market, 2013.
183 Radicati Group, EmailStatistics Report 2019–2023, 2019.
184 Domo, Data Never Sleeps 7.0, 2019.
185 Domo, Data Never Sleeps 7.0, 2019.
186 The categories we present here follow Rosa, *Social acceleration*, 2013.
187 Rosa, *Social acceleration*, 2013.
188 For an empirical exploration of this see Aguiar/Hurst, Measuring trends in leisure, 2007.
189 Becker/Belkin/Tuskey, Killing me softly, 2018; Butts/Becker/Roswell, Hot buttons and time sinks, 2015.
190 See e.g.: Kuss/Griffiths, Social Networking Sites and Addiction, 2017; Song *et al.*, Internet Gratifications and Internet Addiction, 2004.
191 Cheng/Li, Internet addiction prevalence and quality of (real) life, 2014.

5

PRINCIPLES FOR SUSTAINABLE DIGITALIZATION

Everyone's talking about digitalization. Its rapid development seems set to continue in the coming years and decades. But, what will this development look like? That's something we should definitely not leave to chance. On the contrary, we should be consciously and proactively shaping the process. There are already plenty of ideas on how we can ensure that digitalization enhances our individual rights and freedoms – through network neutrality, IT security, data protection and open source, to mention just a few. Thus far, however, the digital and internet policy debates have barely touched on the goal of a social and ecological transition towards sustainability. In the previous chapters, we explored the connections between digitalization, environmental protection and economic justice and suggested ways to introduce a social and ecological dimension. In this chapter, we first summarize the findings of our analyses. We conclude that, although some positive trends towards sustainability are emerging, there are far too many countervailing forces. We then define three guiding principles that we regard as fundamental to achieve sustainable and equitable digitalization. They show how policy-makers, businesses, users and civil society can and should play their part in shaping digitalization to put us on a pathway towards a truly smart and green world.

Summing up: digitalization needs direction

This book's fundamental premise is that our society must transform itself if it is to become sustainable and distribute prosperity more equitably. In theory, digitalization offers numerous opportunities for more eco-friendly production and consumption. It offers starting points for making the way we work, what we produce and who takes the decisions on these matters fairer and more democratic. And, yet, these opportunities are currently being squandered. Indeed, we are seeing too many developments moving into the opposite direction. Under the current legal, economic, political and cultural conditions, social and environmental problems are worsened, not mitigated, by the onward march of digitalization.

In Chapter 2, we described how, in the early days, three distinct and in some ways mutually antagonistic interest groups were instrumental in initiating digitalization: the military, industry and counterculture. For the military, the priority was to use digital technologies to gather more information about – and thus improve surveillance and monitoring of – individuals, organizations and countries. The global corporations' main interest was in using digitalization as a tool to optimize existing fields of business and open up new ones, thereby boosting corporate profits and growth. And those in counterculture who were working for a better world were keen to deploy user-generated, people-centred technologies to enable citizens to liberate themselves and embrace more sustainable lifestyles. In Chapters 3 and 4, we showed how these three conflicting ambitions continue to shape and influence digitalization and its development today. The trends towards excessive government surveillance and commercial data-hoarding and analysis, coupled with intensive digital capitalism, are the main reasons for the failure, thus far, to capitalize on the many opportunities that digitalization affords to improve environmental and social conditions. We believe that policy-makers and society should be doing much more to shape digitalization. They need to adopt a much firmer stance on the direction it should take.

In Chapters 3 and 4, we looked at the mechanics of digitalization and identified three key mechanisms by which it has an impact. All these mechanisms create both opportunities and risks and thus have ambivalent effects on social and environmental sustainability goals. We summarize the inherently ambivalent nature of these mechanisms below. We then consider them in relation to three future-focused guiding principles. Alignment to these guiding principles would unlock the social and environmental potential of digitalization.

So, let's look at each of these mechanisms in turn. First of all, digitalization offers numerous opportunities to improve efficiency and greatly increases the range of options available in many diverse fields. For example, we have shown how e-readers and digital streaming of music and films have the potential to increase energy and resource efficiency compared with analogue alternatives. Online shopping can improve consumption-related energy efficiency, because one commercial vehicle can make multiple deliveries, replacing numerous individual journeys by private car. In the transport sector, traffic management systems based on telematics can boost efficiency, while networked transport planning and sharing schemes help to avoid empty running. And, in industry, manufacturing processes can be digitally optimized so that fewer energy, resource and, above all, labour inputs are needed per unit of output. In addition to this potential for efficiency increases and workforce rationalization, our analyses show how digitalization can broaden the options available in a variety of areas. For example, online shopping is available 24/7 and can be accessed anywhere. Free-floating carsharing schemes are inconceivable without digital tools. Those who have a smarthome system can even use their mobile phone to turn down their heating when they are away from home.

However, our analyses also reveal that switching to digital products and services does not necessarily lead to dematerialization, by which we mean a reduction in the amount of energy and material required for a product or process. First, the development of digital infrastructures and the manufacturing of devices – from e-readers and smart electricity meters to networked industrial facilities – require vast quantities of resources. Such resources have to be extracted, transported and processed. As many of these raw materials are produced under appalling working conditions in countries of the global South, global justice issues come into play. What's more, infrastructures and appliances consume energy, often around the clock, not only in their manufacture, but also when they are in operation. And there is another aspect to be considered: increases in efficiency and options are causing a surge in demand in many sectors. It is precisely because digitalization helps to save materials, costs and time that rebound effects occur. Due to the opportunities afforded by e-commerce, the overall level of consumption within society is increasing. And the manufacturing sector is capitalizing on the opportunities afforded by the industrial internet of things to increase output. On balance, digitalization is at best a zero-sum game for energy and resource consumption: for users, many goods and services are virtual, seemingly dematerialized and resource-light. However, producing and supplying these goods and services is resource-intensive. Increased efficiency and options drive up consumption and, therefore, economic growth, cancelling out the potential savings. And that's not all – the new opportunities are not only boosting consumption; they are also speeding up the pace of life at home and work, making life more stressful for many people.

If digitalization is to support the attainment of sustainability goals, it should be aligned to our first guiding principle, namely **Digital Sufficiency**. We will illustrate precisely what form this guiding principle should take with reference to three further concepts: technical sufficiency, data sufficiency and user sufficiency. But, before we do so, we will briefly describe the other two mechanisms by which digital technologies have an impact.

The second of these mechanisms comprises the sheer volume of information. The new technologies allow data to be collected and shared, while computer-assisted methods offer new opportunities for mass data analysis and use. This information facilitates the global connectivity of people and businesses. It is the basis for energy system transformation, as it supports flexible matching of electricity demand to the fluctuating supply from renewable sources. In the workplace, it creates scope for teleworking and for new and flexible forms of employment such as crowd working. For consumers, digitalization offers a multitude of new information and platforms for both conventional and sustainable consumption, including access to products from all over the world, sharing schemes and marketplaces for pre-owned goods. More generally, information networking offers scope for consumers to become prosumers – whether they are engaged in exchanging or selling hand-made clothing or supply home-generated electricity.

There is a downside to this plethora of information and big data, however: there is a risk of growing government and commercial surveillance. The increasing prevalence of digitalization in the energy supply via smarthome systems etc. raises privacy protection and system stability issues. In the workplace, new forms of monitoring, control and discrimination are emerging. Ultimately, we all risk living completely transparent lives: as workers, as consumers and as citizens. Exposed to the gaze of those who hold our data, we have very little privacy left. The intelligence services – whether in our own country or in far-off dictatorships – can gain access to this data, and companies can use it to perfect their marketing. Personalized advertising and pricing can be used to drive up production and consumption levels, which, from a sustainability perspective, are already far too high.

In order to ensure compatibility with privacy protection, freedom of expression and consumer sovereignty, and to minimize existing power asymmetries between data-gathering organizations and individuals, digitalization must be aligned to the guiding principle of **Strict Data Protection**. Here, too, we define three concepts for more equitable and sustainable digitalization: data sufficiency, privacy by design and the safeguarding of users' data sovereignty.

And, finally, the third key mechanism by which digitalization has an impact is social: it plays a role in wealth redistribution and the polarization of society. Granted, it initially creates new opportunities, enabling more people to gain access to education, consumer choice, political information and so forth, much of which is available free of charge on the internet. It also makes it easier for citizens to make their voices heard and to participate in politics and society via social networks. And, in the economy, digitalization has the potential to establish decentralized networks and make regional or local production more attractive and cost-effective. Overall, it undoubtedly creates opportunities to make the economy and society more open and democratic.

For many people, however, these opportunities currently exist solely on paper. As our analysis has shown, digitalization is polarizing the labour market. Under present conditions, it is leading to the rationalization of jobs, not just in industry, but also in the service sector. Relatively few people are able to find decent work. Some find new jobs, but many more have to contend with precarious and poorly paid employment. The income gap is also widening, with the surging numbers of poorly paid "click workers" contrasting starkly with the dwindling amount of well-paid software developers. There is another driver of this trend: the wage share of national income is shrinking, while income from capital is increasing. Put simply, digitalization is dividing society into digital winners – those who develop the apps and devices and programme or own the robots and shares of IT companies – and digital losers, whose jobs are being replaced and who have no shares in digital businesses.

Digitalization presents opportunities for a decentralized and democratic economy – but, in practice, inequality is increasing in the corporate sector. A

An integrated approach to sufficiency, data protection and the common good will ensure that digitalization makes a positive contribution to social-ecological transformation.

FIGURE 5.1   Principles for sustainable digitalization

Source: Authors' own graphics.

small number of platform operators collect most of the mass data, share most of the advertising revenue, enjoy eye-popping growth in value through rising share prices and often pay extremely low rates of tax. This contrasts sharply with the downward pressure on the rest of the real economy, which is losing market shares while having to shoulder the burden of supporting communities. Digitalization in its current form therefore runs the risk of making society less equitable. In the economy, it is likely to lead to an unprecedented concentration of market power – and, therefore, also political and social power – in the hands of a small number of internet platforms and IT companies.

If digitalization is primarily to benefit, not the corporations and a small minority of individuals, but society as a whole, it must be aligned with the common good. As Figure 5.1 shows, our third principle **Focus on the Common Good** can be broken down into three practical components: the internet as commons, open source and cooperative platforms.

Below, we show how these three guiding principles can help to bring the mega-trend digitalization into alignment with the mega-challenge of social and environmental transition. In addition, we use a backcasting method to develop an ideal social and digital scenario for each principle for the year 2030 (in the boxes). These scenarios entail various practical measures (printed in bold in the boxes). These measures are explained in more detail in Chapter 6.

Guiding principle 1: Digital Sufficiency

The principle of Digital Sufficiency can operate at various levels to curb unsustainable manifestations of resource-intensive digitalization in many areas of life and the economy. It also helps to unlock digitalization's positive environmental potential. The mainstream debate about sufficiency[1] tends to focus on reducing consumption – streaming fewer films, for example, or not buying a new smartphone every two years. Sufficiency can also mean replacing unsustainable with more sustainable behaviours: for example, switching to public transport instead of using private cars. The basic premise of sufficiency is that the problems facing society cannot be solved by new technologies alone. Behavioural changes are also required to create synergies. For example, we can use digital opportunities either to replace conventional vehicles with self-driving cars or to promote smart public mobility. The latter option is more sustainable from an environmental perspective, but would require many people to change their behaviour.

We are currently witnessing the incursion of digital technologies into almost every area of life and the economy – in the workplace, in leisure and recreation, at home, in dating, vacationing, education, politics, the financial markets and consumption. Viewed objectively, it clearly makes no sense to deploy one set of technologies in an effort to satisfy every conceivable social need or solve every problem. "One size fits all" solutions have often caused problems in the past. The phase-out of local public transport in many US cities with the advent of the automobile in the first half of the 20th century is a case in point. In our own era, there needs to be a critical social debate about where digitalization makes sense and where it does not. As we have seen, there are major advantages in having a degree of digitalization in the energy system. But, this certainly does not mean that it would be a good idea to amplify all our domestic appliances and devices by energy- and data-intensive smarthome systems.

So, we need to engage constructively and critically with digital technologies and applications. At present, the maxim appears to be "as much digitalization as possible". We offer an alternative: we believe in "as much digitalization as necessary and as little as possible". If we wish to avoid transgressing planetary boundaries, there is no other option but to substantially reduce our energy and resource consumption, including that due to digitalization. We suggest three approaches to operationalize the guiding principle of digital sufficiency.

(1) **Technical sufficiency** means that information and communication systems are designed in such a way that few devices are needed and rarely have to be replaced. In most cases, this will require joined-up thinking about hardware and software. At present, the introduction of increasingly data-intensive software leads to hardware – laptops, smartphones, etc. – that is still in perfect working order having to be replaced with newer models. There needs to be an emphasis on product longevity in hardware and software development. Technical sufficiency also means making efforts to ensure that manufacturing is sustainable and equitable, with a focus on devices' repairability and modular expandability. Every digital device should, as far as possible, have eco-friendly design built in.[2]

As the Fairphone shows, there is potential to design smartphones in such a way that components, such as the screen and the battery, can be replaced, extra memory added or processing power increased.[3] The shift towards smart appliances – such as refrigeration/cooling systems and washing machines – as part of a decentralized energy system transformation offers scope for solutions to upgrade existing devices. This removes the need to replace older machines with new ones. Open source has a key role to play in ensuring the technical sufficiency of hardware and software.[4] If technical design information is widely accessible, maintenance and repair become much more straightforward and product lifespan can be extended more easily. Open source also creates scope to introduce a wide range of durable products manufactured by persons other than the designers and patent holders.

(2) **Data sufficiency** relates to the design of digital applications. More data traffic requires more server capacity and IT infrastructure. The way in which software has been developed over the years has often resulted in an increase in the volume of data traffic – much of it related to ancillary and non-essential background services.[5] Many apps are continuously accessing clouds, but would actually work perfectly well offline with occasional data updates. In Chapter 3, we provided some examples relating to media consumption. There is no need for constant streaming of songs that the user would like to listen to multiple times; instead, they can be downloaded once and stored on a local device. Video portals should at least make users aware of the option of lower graphic resolution; better still, they could offer the lowest resolution as the default setting. For these and other applications, the principle of *sufficiency by default* should apply. These may be small steps, but every little helps. If there are, indeed, millions of self-driving cars on our roads in the future, this will create a new set of demands on our digital infrastructures and further increase the ecological footprint of digitalization. Data sufficiency raises one key question: how much permanent connectivity and data traffic is useful and necessary? Every discussion about the smart city, smarthome, smart mobility or the internet of things should start with this critical question. The answer, in every case, is this: the smaller the data volume, the less need there is to expand resource-intensive infrastructures or manufacture increasingly powerful devices and server farms. Data sufficiency helps to ensure that infrastructural expansion remains modest in scale, thereby conserving resources every step of the way, from broadband rollout and mobile communications standards to data centre construction. This would minimize energy consumption even for complex digital applications.

(3) **User sufficiency**, lastly, is based on the recognition that sustainability goals cannot be achieved with smart technology alone: it is essential to change mindsets and user behaviour as well. If smartphones break, users can try and have it repaired instead of immediately buying a new one – provided that technically sufficient design allows this. The internet creates opportunities to buy clothing,

appliances and furniture second-hand instead of new. Smart grids make a decentralized energy system technically feasible. Yet, the transformation of the energy system is also based on local initiatives and engagement by individuals working on the ground. The main purpose of user sufficiency is to ensure that digital tools do not lead to rebound effects and increased consumption. Smart networking in the transport sector makes it possible to travel from A to B more quickly and cheaply. Yet, sufficiency-oriented users will not see this as an opportunity to travel more. And, if digital communications mean that work and logistics can be managed more easily, sufficiency-minded users will not increase their activity, but will enjoy the fact that they have more free time. Ultimately, every user must ask themselves this: how many digital devices and how much permanent connectivity do I need for a good life? The ongoing costs and disadvantages of our "imperial mode of living"[6] are ultimately borne by other people, often in the global South, and by the environment. Those impacts underscore the importance of user sufficiency.

Social utopia: digitalization goes green

Let's cast our minds forward to 2030. In this scenario, we are living in a digitally sufficient world. How did we achieve it? We did so primarily by avoiding anything that leads to hyper-consumption in the digital space. With that aim in mind, policy-makers – in response to growing **debates on digitalization** – introduced **advertising bans** in many areas of the internet and imposed a universal **passivity rule** on companies to prevent them from manipulating consumer preferences. Hardware, such as smartphones and computers, is built modular, repairable and open source, remains in service for as long as possible and all applications are designed to be data-sufficient, regulated by an **ICT design directive** and incentivized by a **digital-ecological tax reform**. At the same time, **critical education programmes** have shown us how to use digital options wherever they can help to minimize harm to the environment. For example, we avoid business travel by meeting in the virtual space; personal transport is based on shared use; and material consumption needs are satisfied largely through **peer-to-peer sharing** schemes. The state has intensively promoted **digitalization in selected sectors** such as local public transport. And we have re-regionalized business practices and thus reduced the distances travelled. In this more localized economy, people use digital technologies to support consumption of **sustainable products** and create **social innovations** that promote sustainability. Digitalization has thus made a contribution to halting climate change, while promoting lifestyles and business models that take account of the needs of future generations.

Guiding principle 2: Strict Data Protection

With the increased use of the internet and digital applications, people and companies are revealing more and more information about themselves. Many applications exist solely on the basis of this data – including some that are socially and environmentally beneficial, such as decentralized micro-grids in the energy market, ride-sharing apps in the transport sector and prosumer platforms for local commerce. Decentralized regional business networks connect stakeholders who are dispersed across a wide area. However, alongside these opportunities, there are also risks: the state and some monopolistic corporations use this mass data to encroach on users' privacy and monitor their activities. Every market player with access to personal data must therefore be bound by strict privacy rules. As we will show, this is vital for two main reasons: to safeguard democracy and to avoid transgression of planetary boundaries.

First of all, strict data protection is essential in the interests of society at large to safeguard privacy, personal integrity and freedom of opinion. These are important objectives in themselves and are recognized by the international community as inalienable core human rights in the International Covenant on Civil and Political Rights, adopted in 1966.[7] Privacy and freedom of opinion are prerequisites for a well-functioning democracy. In the political sphere, citizens can only form and express their own independent opinions if they are not monitored and have no cause to fear reprisals at the hands of the state. Even now, the fear that political conditions could change in future and that stored data might have repercussions are inhibiting freedom of expression and democracy. Furthermore, large companies can misuse big data in order to manipulate citizens' political views and influence public debate. There is an excessive concentration of data in the hands of a relatively small number of global corporations such as Alphabet, Facebook, Apple, Microsoft and Amazon, and the data analytics firms that work for them. This is at odds with a well-functioning democracy, as it enables opinions to be manipulated and undermines individual autonomy. Strict limits on data retention by governments and the private sector are essential to prevent the erosion of democracy. The slogan here is: Whose data? Our data!

Second, strict data protection is indeed important for the environment. Currently, the corporations mentioned above collect and analyse data primarily for commercial purposes. Their main objective is to use personalized advertising and pricing or situational marketing, based on movement profiles, to increase the already high and unsustainable level of consumption. This can be prevented by a comprehensive approach to data protection, which would also contribute to key sustainability objectives, notably the Sustainable Development Goals adopted by the United Nations[8] and the Paris Agreement on climate change.[9] Commercial data analysis and use must therefore be regulated far more stringently than before. Moreover, the principle of strict data protection should not be undermined on social or environmental pretexts. Smarthome systems, bikesharing, carsharing and the movement profiles produced for digitally connected local public transport should not allow conclusions to be drawn about individual users. There is a substantial risk that digital applications would otherwise proliferate to the point where they facilitate commercial manipulation or government surveillance of citizens. There are three approaches to implement the principle of strict data protection:

(1) **Data sufficiency**, which is already set out above, is a matter of ensuring that as little information as possible is generated and uploaded to the cloud or transferred between data centres and providers. Data sufficiency not only conserves natural resources and saves energy; it also supports data protection. Here, the principle of data sufficiency overlaps with the established principle of data minimization:[10] the less data is generated and transferred, the fewer opportunities arise to misuse the data, encroach on privacy, influence consumers and subject citizens to surveillance. Data sufficiency is thus a key element of strict data protection.

(2) **Privacy by design** means ensuring that devices and applications always guarantee a maximum level of privacy protection. This means that operating systems, browsers, apps, routers, etc. should be designed with data anonymization as standard. Privacy by design was recognized as an essential component of privacy protection at the International Conference of Data Protection and Privacy Commissioners in 2010. It is governed by the provisions of the EU's General Data Protection Regulation, adopted in 2016.[11] However, privacy by design has so far focused mainly on data security rather than on strict data protection. The latter is not only about ensuring that data does not fall into the wrong hands; it is about preventing its collection in the first place.[12] In order to advance strict data protection, privacy by design must be adopted as the global standard.

(3) **No data commerce:** Binding and comprehensive data protection rules need to be rigorously upheld by the private sector. It is essential to close the gap between the extensive data protection rules that apply to public institutions, such as governments and intelligence services, at least in several countries including those of the EU, and more lenient privacy rules applicable to the private sector. In those countries, the public sector is governed by stringent rules on data retention and foreign-to-foreign telecommunication surveillance.[13] By comparison, the rules applicable to the private sector are extremely lax. The Act on Airline Passenger Data Storage in Europe is a case in point: it strictly governs the storage of data and its management by the German Federal Criminal Police Office, but these rules do not apply to the airlines themselves.[14] Hence, if you book a plane ticket, you have to accept that the airline will keep your passenger information on record for several years. It is a similar situation in other areas: if you wish to use the Google search engine, you must first consent to its terms of service and allow cookies to be stored on your computer and all your search items to be stored pretty much forever – and used for various purposes by Google. If you set up a LinkedIn account, you tick a box to show that you accept the terms and conditions. Most users have no idea what they are allowing their data to be used for and which rights they are relinquishing. They do not realize, for example, that by agreeing to its terms of service, they grant Facebook "a non-exclusive, transferable,

sub-licensable, royalty-free, and worldwide license" to use their content.[15] In effect, they are allowing the company to use all their uploaded photos and information for its own commercial purposes. Users of these quasi-monopoly internet platforms, search engines and social networks have no opportunity to protect their data effectively, other than by opting out of social media altogether. Anyone who does not want to stop using WhatsApp, perhaps for fear of being socially excluded, and who has not managed to persuade their friendship group to switch to other more secure services, will have no real control over the extent to which their data is disclosed to Facebook. Voluntary certification of services that are trustworthy and committed to data minimization – such as the EuroPriSe Privacy Seal – is a first step at best.[16] This is not the way to achieve strict data protection or to curb the relentless exploitation of private information by business cases that proliferate a kind of "surveillance capitalism".[17] There must be strict limits and clear rules on the collection, storage and use of data and on data-sharing with third parties.

Social utopia: people own their data

Let's cast our mind forward to 2030 and imagine a world in which our data belongs to us – to each and every individual. The risks of government surveillance and commercial manipulation that were rife in the 2010s and 2020s are largely a thing of the past. The digital technologies that dominate our lives are now developed in strict compliance with the privacy by design principle. Strict data protection is, for the most part, a reality – partly as a result of pressure from a **broad civil society movement**. No one has access to our data without our active consent. How has this been achieved? The first step was to enforce and implement full stakeholder compliance with existing data protection laws, e.g. on **data minimization** and a **coupling ban**. A selective **advertising ban** has done much to restrict personalized online advertising. An **Algorithm Act** guarantees that no one needs to worry that they will no longer have access to credit from their bank simply because of their online social media profile. And, thanks to a **passivity rule**, there is no need to be concerned that a link might lead to hidden advertising rather than genuine information. In other words, policy-makers have put in place many of the conditions required for strict data protection. However, users too have played their part – by switching to **sustainable digital products**, by making an informed choice not to use data hydras and by opting instead for secure, data-sufficient alternatives. Civil society has made a major contribution by providing **critical digital education** and **shaping debates on digitalization**. Taken together, these measures have helped to ensure that, in 2030, we can still exercise our right of self-determination in a democratic society.

Guiding principle 3: Focus on the Common Good

As we have shown, under current conditions, the way in which the benefits arising from digitalization are shared is extremely inequitable. In the labour market,
only a relatively small number of workers with appropriate skills can count on job
security and rising incomes. Owners of capital assets are seizing control of much
of the value added that is generated by highly automated production. Meanwhile,
large internet platforms and IT companies make virtually no financial contribution to the common good. It seems that, at present, the maxim is "profit first,
public welfare second".

This has to change. If digitalization were consistently aligned to the common
good as a guiding principle, far more people – indeed, society as a whole – would
benefit. One debate that is already focused on the common good concerns the
issue of network neutrality.[18] Net neutrality means that the internet is equally
accessible to all providers and users and does not discriminate by content. Without net neutrality, peer-to-peer models such as BitTorrent, which enables sharing
of files and data between users, would operate much more slowly. The numerous
and frequent attacks on net neutrality by corporations and legislators are aimed
at entrenching power, monopolies and existing inequalities.[19] They also prevent
new and innovative applications and start-ups from becoming established. Fright-
ening, too, is that about half of the world's data transmission capacity through
glass fibre cables in the oceans are owned by Alphabet, Amazon, Facebook and
Microsoft. There is increasingly less separation between those who provide content and those who run the infrastructure.[20]

By contrast, a genuine focus on the common good would not only create
a level playing field on the internet in technical terms. It would also make the
digitalization of substantial sectors of the economy more equitable. Above all,
clearly defined policy frameworks are needed to ensure the equitable distribution
of employment, incomes and power. So, what action can policy-makers take
in relation to the labour market, tax law and energy prices, for example? We
provide some answers in Chapter 6. In addition, the digital contribution to a
social-ecological transformation crucially depends on who develops and shapes
the process of digitalization. In recent years, this has increasingly been dominated
by major corporations, so it is hardly surprising that their owners and employees
are currently the main beneficiaries. If digitalization is to focus on the common
good, digitalization must not only be done differently: it must also be designed by
different stakeholders. In the next section, we outline three digitalization strategies that focus on people, not profit.

(1) **The internet as a commons:** The internet is a textbook example of a
virtual commons. In the Northern hemisphere, the concept of "commons" dates
back to the Middle Ages, when peasants owned pasture collectively and took
turns to graze their animals. But, indeed, countless examples of successful local
and global commons exist today.[21] The internet displays all the key characteristics

of a "commons": it exists solely because users themselves have created it and are continuously developing new versions of it. Internet content is available to everyone; no one is excluded. There is rarely any competition over access to content. And, perhaps most importantly, the internet flourishes best when it is not dominated by singular (private) interests. However, as we have already said, a small number of monopolies are now colonizing both users and content for their own purposes. The internet is thus at risk of mutating: away from being a virtual commons where all users – whether commercial, civil society or private – can interact on equal terms, and towards becoming a strongly capitalist, almost neo-feudal marketplace dominated by a small number of players.[22] There is already considerable and worrying inequality between the small number of providers and the masses of users, with the latter increasingly being thrust into the role of more or less passive consumers. In order to change this situation and ensure that emphasis is placed on the common good, the internet should be restored to its rightful status as a commons, with users taking priority. They should have an active role and a genuinely free choice as to which information, services and products they wish to contribute and use. Social and environmental preferences will only come to the fore in the digital space if there is a ban on the subtle inducements resulting from the use of anonymous commercial bots, personalized advertising, interest-led rankings of search results and the application of selective algorithms. Platforms such as Wikipedia, one of the few large-scale commons-based platforms on the internet, are a cause for hope. However, they must cease to be the exception and become the rule if the internet is to focus once more on the common good.

(2) **Open source** means that human knowledge is publicly accessible, for everyone's benefit. Open source is deeply rooted in the history of digitalization and the internet. As we saw in Chapter 4, there are still some examples of success-ful open-source applications – but they are rarely market leaders. We have already shown, with reference to technical sufficiency, that open source is an important element in achieving eco-friendly product design and modularity, ensuring that digital devices can be repaired and updated. Open source also helps to promote a focus on the common good, given that open-source and free software such as the Ubuntu operating system, the private messenger service Signal and Libre-Office software are generally available at no charge. Open-source hardware, such as the products sold by Adafruit Industries and SparkFun Electronics, is usually a low-cost option and can thus be accessed by almost any internet user. Open source also enables users to carry out their own DIY repairs and maintenance, so that fewer services have to be bought in the marketplace. It can also be a starting point for users to develop their own business models based on publicly acces-sible designs, thus helping to build a more democratic economy. Broad-based application of the open-source principle in software and hardware development would radically transform many business segments associated with digitalization and promote a focus on the common good.

(3) **Cooperative platforms** support fair decision-making and equitable distribution of profits in the digital economy. As we have shown, the formation of monopolies on the internet is partly the outcome of network effects, whereby one or a handful of global platforms – Facebook, Google Plus, LinkedIn and the like – rather than a large number of small local applications come to dominate. In some sectors of the digital economy, we have a situation that economists would call a "natural monopoly". In these cases, having many small-scale and widely dispersed providers makes little sense. However, it should be possible to distribute the platforms' income and power more equitably within society, even if natural monopolies exist. This is where cooperative platforms come in. They can take a variety of forms,[23] with support coming from members, communities, cooperatives or trade unions. Their chief characteristics are common ownership, democratic codetermination and fair distribution of revenue. Wherever internet platforms are created, "collaborative, not capitalist" should be the watchword.

Social utopia: everyone benefits from digitalization

Let's cast our minds forward to 2030 and imagine a world in which everyone benefits from digitalization and has a hand in shaping it. Unlike the early days of digitalization at the start of the millennium, this world is no longer dominated by a handful of highly profitable corporations that decide more or less on their own terms what a digital society should look like. Policy-makers have taken action to curb the concerted financial might of the digital giants and their well-resourced legal and lobbying departments by introducing a **reform of monopoly law**. The monopolies of the early days no longer exist – for policy-makers have also **strengthened platform cooperatives**. As a result, the digital economy has changed to such an extent that one of its predominant features is now democratic decision-making. Users have access to **critical digital education** and adopt a constructive approach to digital opportunities. They rely primarily on **peer-to-peer sharing** schemes and are engaged as prosumers, with the result that many areas of consumption are now organized collaboratively. The excesses of globalization have given way to a phase of **re-regionalization**, which has restored the balance between global and local wealth generation and **slowed the pace of life**. The improvements in productivity made possible by digital technologies have not led to unemployment on one side and wealth on the other, as happened in the past. Instead, employment is now shared more equitably, not least through **shorter full-time working hours** for everyone. This frees up capacities that can be used to **expand the care economy**. Digitalization has thus helped to overcome the major challenge of demographic change and safeguards peaceful social relations.

Commons, open source and cooperative platforms are complementary, not conflicting concepts. Like the commons, many people contribute to and benefit from open-source applications. And open-source technologies are ideal for application in an internet as a commons and on cooperative platforms.

Looking at the three guiding principles of Digital Sufficiency, Strict Data Protection and Focus on the Common Good in unison, it is clear that these principles have major implications for business models and, hence, for the structure of the entire digital economy. Granted, they conflict in quite fundamental ways with the current logic driving most internet and IT corporations. However, change is needed – not only from the providers, but also from the users of digital devices and applications. Products with a long service life not only thwart efforts to generate a quick profit; they also require a shift away from "fast fashion" and a throw-away mentality in the digital sphere. Overall, it is clear that a consistent alignment of digitalization to these three guiding principles will not come about of its own accord, but must be demanded, promoted and implemented systematically by policy-makers, civil society and users. Progressive start-ups and companies must also be given targeted support to ensure that they participate in this mammoth task through smart collaboration. Some of the practical policies, measures and initiatives that we believe make sense in effecting this sustainable digital transformation are described in the next chapter.

Notes

1 Alcott, The sufficiency strategy, 2008; Gorge *et al.*, What Do We Really Need?, 2015; Gossen/Ziesemer/Schrader, Why and how commercial marketing should promote sufficient consumption, 2019; Sachs, Die vier E's, 1993; Princen, Principles for Sustainability, 2003; Princen, *The logic of sufficiency*, 2005; Schneidewind *et al.*, *Economy of sufficiency*, 2013.
2 Hilty *et al.*, Green Software, 2015; Joumaa/Kadry, Green IT, 2012; Kern *et al.*, Sustainable software products, 2018; Naumann *et al.*, The GREENSOFT Model, 2011.
3 Proske/Jaeger-Erben, Decreasing obsolescence with modular smartphones?, 2019.
4 Gibb, *Building open source hardware*, 2014; Hilty *et al.*, Green Software, 2015; Wittbrodt *et al.*, Life-cycle economic analysis of distributed manufacturing with open-source 3-D printers, 2013.
5 Hilty *et al.*, Green Software, 2015.
6 On the imperial mode of living see Brand/Wissen, *The limits to capitalist nature*, 2018; Kopp *et al.*, *At the Expense of Others?*, 2019; for an earlier discussion see also Sachs/Santarius, *Fair future*, 2007.
7 United Nations, *International Covenant on Civil and Political Rights*, 1966.
8 United Nations, *Transforming our world: the 2030 Agenda for Sustainable Development*, 2015.
9 UNFCCC, Paris Agreement, 2015.
10 See e.g. European Commission, *Regulation of the European Parliament and of the Council on the protection of natural persons with regard to the processing of personal data and on the free movement of such data (General Data Protection Regulation)*, 2016.
11 European Commission, *Regulation of the European Parliament and of the Council on the protection of natural persons with regard to the processing of personal data and on the free movement of such data (General Data Protection Regulation)*, 2016.

12 Schaar, Privacy by Design, 2010.
13 For instance, Daten-Speicherung.de, Minimum Data, Maximum Privacy, 2017.
14 European Union, *Directive (EU) 2016/681 of the European Parliament and of the Council of 27 April 2016 on the use of passenger name record (PNR) data for the prevention, detection, investigation and prosecution of terrorist offences and serious crime*, 2016.
15 See Facebook, *Terms of Service*, 2019.
16 Christl, Corporate Surveillance In Everyday Life, 2017.
17 Zuboff, *The age of surveillance capitalism*, 2019.
18 Wu, A Proposal for Network Neutrality, 2002.
19 For an overview see e.g. European Digital Rights (EDRi), Net Neutrality, 2019.
20 Satariano *et al.*, How the Internet Travels Across Oceans, 2019.
21 Bollier/Helfrich, *Free, fair, and alive*, 2019; Ostrom, Governing the Commons, 1990.
22 See on this e.g. also Berners-Lee, The past, present and future, 2016; Morozov, *The net delusion*, 2011.
23 See e.g. Scholz, Platform Cooperativism, 2016.

6

AGENDA FOR A NETWORK SOCIETY

The guiding principles that we have described will not become reality of their own accord. We shall wait in vain to see them put into practice unless they are pursued actively – of course by businesses, through leadership and progressive corporate responsibility measures, but also by policy-makers, through appropriate incentives and regulation, by users, through their consumption behaviour, and by a critical civil society that pays much more attention to digitalization issues than it has done in the past. In this chapter, we shall highlight how politicians, users and civil society can contribute to social-ecological digitalization. However, there is no master plan that will ensure that digitalization proceeds along equitable and sustainable lines. The agenda for a connected, yet sustainable, society must therefore be framed one step at a time. The following suggestions are intended to stimulate further discussion and encourage people and organizations to initiate practical action.

Signposts for a transformative digital policy

For many politicians the world over, digitalization is now high on the political agenda. Dozens of local governments worldwide call for smart city strategies. At the country level, governments pursue strategies for the digitalization of economy, infrastructures (e.g. broadband, 5G networks etc.) or public institutions. For instance, Germany leverages investments both into Industry 4.0 as well as into "AI made in Germany". At European level, the issue was already addressed in the "Digital Agenda for Europe – Driving European growth digitally",[1] launched in 2010. Yet, as the title of the early European Agenda reflects, so far the majority of political programmes focus on economic goals. Growth and employment are top priorities – only sometimes followed by "access and participation" and "trust and security". Goals such as resource protection, energy efficiency or social cohesion are rarely mentioned.

There is little sign of policy-makers taking a proactive role in shaping the course of digitalization towards fostering the larger common good. A transformative political vision that joins up thinking about digitalization and global sustainable development is currently absent. Political institutions – such as national governments, but also local authorities, public bodies and government agencies – need to become far more engaged. And they need to be quick about it. With each year of inaction that passes, it becomes more difficult to establish conditions that will steer digitalization in a socially and environmentally sustainable direction. How can the state (re-)gain political primacy?

Some of the elements of a transformative digital policy are outlined below. The path is always defined by the three guiding principles of Digital Sufficiency, Strict Data Protection and a Focus on the Common Good. The first aspect of digital policy that we consider involves economic "rules of engagement" that define the framework for the digital economy. Second, we show how the state can proactively support social-ecological digitalization by creating incentives. Third, policy-makers need to put in place ancillary measures to embed digitalization in a setting that is designed to promote environmental sustainability and the common good.

Legislate an ICT design directive

The use of digital services is by no means virtual, but involves significant amounts of physical resources and energy throughout the entire lifecycle. To begin with, the mining of raw materials for digital devices is oftentimes characterized by insufficient environmental standards, as well as low wages and poor safety conditions for the workers, particularly in countries of the global South. As we have shown in Chapters 3 and 4, the production process involves toxic chemicals and is highly energy intensive, while working conditions in the digital sweatshops of Foxconn and other factories usually do not comply with core labour rights. In the use phase, consumer devices as well as data centres require electricity and account for greenhouse gas emissions. And, at the end of the lifecycle, only a miniscule share of digital devices is properly recycled.

Governments should develop ICT design directives that require minimum social and environmental standards through the entire production chain. Furthermore, legislation can help to extend the useful lifetime of hardware, which reduces the demand for new devices and slows down the flow of resources from extraction to waste. For instance, rules can require manufacturers to avoid critical resources, use recycled materials, guarantee longer warranty periods and adopt minimum standards for modularity, repairability and upgradability of devices. Manufacturers should be obliged to provide software updates for operating systems until the end of the physical lifetime of devices. Mandatory take-back programmes for retailers can increase recycling rates. Furthermore, governments should set standards, not for relative energy consumption (efficiency), but for absolute energy consumption of devices (sufficiency) as well as for data centres, which should be run on 100 per cent renewable energies. New data centres should be smartly integrated into urban dwellings to make best use of waste heat. And, finally, government procurement can support open source hard- and software, creating markets that are more democratic and empower DIY.

Introduce selective advertising bans

The internet is increasingly becoming the principal forum for economic exchange and social dialogue. This space must be clearly and consciously shaped by policy-makers. The web is currently full of personalized advertising and other elements that are hostile to democracy and boost consumption (see Chapters 3 and 4). This needs to change. We support ad-free spaces on the internet, particularly on search engines and social media. Digital sufficiency and the internet as commons are incompatible with ubiquitous advertising. To enable the internet to move further towards being a place for self-determined consumption and peer-to-peer social dialogue, we would like to see selective advertising bans online. This, of course, affects the core business of some of the world's largest corporations, notably Alphabet and Facebook. However, fears that their services could not exist without advertising are exaggerated. Instead, we need to ask ourselves whether we really want to pay for the use of search engines, messaging services and social media with our personal information rather than with money. There are already alternatives to services such as WhatsApp, including Threema, Signal, Telegram or Wire, which provide users with more secure communication services without an advertisement-driven business model. Similar offers exist for search engines and social media. The advertising industry will vigorously resist a partial advertising ban on the internet. But, the longer policy-makers wait, the more powerful the opposition will become and the more difficult it will be to bring forward the necessary legislation.

Develop a passivity rule

As a second measure to promote digital sufficiency and a focus on the common good, we are calling for a more general passivity rule. Every stakeholder, whether commercial or non-commercial, must refrain from engaging in any practice that is intended to manipulate users. How can policy-makers implement such a rule? Let's consider some examples. Bots – software applications that perform tasks on a largely automated basis – are increasingly programmed to imitate human behaviour. They are used to influence opinions on social media via posts or tweets or to rate and advertise products and services – always in the guise of supposedly authentic consumers. There are already bots that specialize in collecting large quantities of data about particular individuals with the deliberate intention of exerting influence over them.[2] If policies dictate that bots must always be labelled as such, users will be less at risk from this subtle manipulation. Another example of influence is the personalized provision of information by online platforms on the basis of certain criteria, such as age, place of residence, usage patterns or preferences. The passivity rule would stipulate that providers must always reveal what data is being used as a basis for the provision of information. All users should be able to change these criteria at any time. The passivity rule goes hand in hand with the selective advertising ban: major restrictions must be placed on any data collection and analysis aimed at boosting consumption via personalized advertising. Like advertising bans, a passivity rule would encourage sufficiency as a behavioural pattern, do more to align the internet to the common good – and represent a major step towards more data privacy.

Pursue data minimization and a ban on data coupling

Legislation restricting the escalating collection of data would do even more to further the implementation of the guiding principle of strict data protection. A study that has analysed traffic from more than 5,000 apps used by thousands of users in manifold countries discovered that more than 70 per cent of the analysed apps dedicate at least 10 per cent of their traffic to tracking and advertising activities.[3] The principle of data sufficiency (data minimization) is already enshrined in privacy legislation of some countries: it means that personal data may only be collected if it is genuinely needed for a specific application.[4] However, even where implemented, the practical implementation of this principle leaves much to be desired. If search engines, social networks, map services and other apps no longer collect masses of data because they are banned from doing so, they will no longer be in a position to misuse it.

Another aspect involves the sharing of data – currently a widespread practice in the digital economy and one that makes it almost impossible for users to know who is doing what with their personal information. In theory, the "coupling ban" prevents data being passed on unless users have voluntarily given their consent.[5] The problem is with the word "voluntarily": users are often faced with the choice of either consenting or not being able to use the application at all.[6] The logical extension of the coupling ban would involve prohibiting any sale or exchange of users' personal data between private businesses.

As with an advertising ban, enforcement of the principle of data sufficiency and extension of the coupling ban might result in businesses losing some of their revenue. Some applications that we have until now "paid for" with our personal information might then become more expensive or become chargeable instead of free. Concepts of "data sovereignty" are based on the premise that users themselves should decide whether they want to pay to use an application by releasing their data or by some other means. However, there will be people who, for financial reasons, have no real freedom of choice or who feel obliged to participate in social media or other activities for social reasons. An even more serious aspect is the fact that vast numbers of people – probably hundreds of millions – are passing on, not only their own data, but also that of their contacts, for example when those are accessed by smartphone apps. The UK's data protection watchdog – the Information Commissioner's Office (ICO) has already ruled that this is an infringement of existing privacy rights.[7] Various other European countries have also raised concerns over data sharing, including France, which ordered WhatsApp to stop sharing data in December 2017. In the EU, the key legal and regulatory issues relevant to the processing of personal data by mobile apps stem from the General Data Protection Regulation (GDPR).[8] The EU Member States recently pushed a crucial reform on privacy norms – the ePrivacy Regulation, an update to the outdated 2002 ePrivacy Directive – close to a dead end.[9] In the USA, users who suspect a site or an app mishandles their personal information can file a complaint with the Federal State Commission (FTC). The FTC does not resolve individual complaints, but such complaints may contribute to an investigation or enforcement action.[10] Hence, individuals have legal protection, but privacy laws have not kept pace with technology. Governments need to adopt and, in particular, enforce standardized legislation to guarantee strict data protection.

Draw up an Algorithm Act

A passivity rule and advertising bans aim to prevent subtle manipulation. In addition, algorithms are being used to drive automated decision-making systems that cannot and are not intended to be uninfluential – in particular, in the case of machine-learning algorithms (artificial intelligence). Here, too, we need to define what criteria the algorithms use as a basis for decision-making, what the aims of algorithm use are and what values underpin the process. The use of algorithms and the foreseeable developments as regards self-learning algorithms in a world that is becoming increasingly interconnected all affect deep-seated ethical principles of our communal life. It is up to politics to face up to this challenge. Leaving things to business, as in the past, is not a serious option.[11]

As a first step, companies that develop automated decision-making systems should be required to reveal all the algorithm's criteria underlying its decision-making. The principle of strict data protection must apply here too. For example, it may be appropriate for an app designed to increase the use of public transport to offer automated personalized suggestions for certain forms of mobility. After all, what is the point of offering a bicycle to a person in a frail state of health or a hire car to someone who does not have a driving licence? But, users must always know what these criteria are and must be able to change them, so that they are not simply passive recipients of the algorithm's decisions.[12]

The Algorithm Act could also help to ensure that software takes account of normative issues that serve the common good.[13] To cite another transport-related example: for environmental reasons, it may be appropriate to prioritize bikesharing offers over carsharing or taxis. Policy-makers will also need to discuss which normative criteria should serve as a basis for decision-making by the algorithms used in self-driving cars in certain situations. In addition to ethical principles, social-ecological objectives could have a bearing here. For example, the algorithms for self-driving minibuses could be programmed to avoid urban routes that are already served by mass transport systems, such as buses, rapid transit rail and the underground.

Reform monopoly law

It is said that data is the oil of the 21st century.[14] And today's policy-makers can learn from how the oil industry was handled in its early beginnings. The spectacular breakup of the Standard Oil Company by John D. Rockefeller under President Theodore Roosevelt early in the 20th century was an important step in safeguarding democracy and competition and one that was followed elsewhere in the oil industry by other cases under monopoly law. A growing number of voices are now calling for a similar breakup of the major digital corporations.[15] As we have seen in Chapter 4, today's internet giants have vast capital assets. And, because of the amount of information they possess, they also have immense social and political power that far outstrips that of Standard Oil a century ago.

If digitalization is to be based on democratic and ecological criteria, the primary justification for regulation of dominant companies through monopoly law is the need to maintain policy-makers' capacity to act. The more powerful individual corporations become, the more difficult it is for policy-makers to regulate them. In addition, preventing oligopolies and monopolies on the internet is necessary from a consumer protection perspective, particularly in order to implement the guiding principle of strict data protection. This is essential if excessively large power asymmetries between consumers and providers are to be prevented. A third reason for taking action under monopoly law is to safeguard competition, innovation and fairness in the internet economy. In terms of commercial competition, the concentration of vast quantities of data and knowledge in the hands of a few market leaders poses an almost insurmountable barrier to start-ups, preventing them from equal participation in the development of forward-looking technologies, such as artificial intelligence.

In the light of these important considerations, there are calls for platforms larger than a certain size to be either expropriated or at least declared "public institutions".[16] In the case of companies whose market penetration is on the scale of that of Facebook, it could be argued that they should no longer be treated as private or public limited companies: their influence is so strong that users who are excluded or censored have virtually no opportunity to express their opinions or make their voices heard elsewhere. There should not be areas in which profit-oriented companies can "eavesdrop" on and potentially manipulate the public and private conversations of billions of people: these areas should be taken over either by the public sector or by democratically organized non-profit institutions (such as, for instance, public foundations). Most importantly, any concentration of data and knowledge should be systematically scrutinized in the light of cartel law.

Strengthen platform cooperatives

Policy-makers are not restricted to establishing meaningful rules of engagement in the form of laws, rules or bans for the internet – they can also actively promote the social-ecological orientation of digitalization. This is readily achievable, both for platforms regulated by monopoly law and for those that are cooperatively organized. Many of the major internet companies – Instagram, Snapchat, LinkedIn, Uber and others – are platforms or operate platforms. They produce no products, but act as intermediaries and hence lay down the rules for all users of these platforms. As we have seen, network effects and high start-up costs mean that platforms tend to become monopolies; it is, in effect, a "winner-takes-all" scenario. Their market power enables these corporations to rake in high profits – in part because they often have low wage costs and pay very little tax. In recent years, platform cooperatives have emerged as an alternative model that contrasts with the prevailing "platform capitalism". "Platform cooperativism" mimics the concept of the existing capitalist platforms, but transfers it into the context of the solidarity economy.[17] In this new system, innovations no longer serve the purpose of proprietary profit maximization, but, instead, further the common good.

Platform cooperatives can be organized in various ways. The most widespread form is the cooperative. For example, the online marketplace Fairmondo provides a social-ecological alternative to Amazon. It focuses on fair and sustainable products, including used goods, and is managed cooperatively by its sellers and customers. On Loconomics, independent service professionals have joined together to offer an alternative to proprietary sites such as Taskrabbit and MyHammer. These independent service providers own shares in Loconomics, receive dividends – if there are any profits – and make their voice heard. The findings of a case study comparing the capitalist platform TaskRabbit and the platform cooperative Loconomics confirmed that the democratic practices of Loconomics allow it to better satisfy workers' needs compared to platform capitalist enterprises.[18] "Produser" platforms, like Resonate (music), Members Media (film) and Stocksy (stock photography) represent another model of communal organization. Ownership is shared between producers and users. On Stocksy, photographers can sell their images and receive a share of the platform's profits in addition to their fee. On streaming website Resonate, listeners pay each time they listen to a song; if they listen to a particular song often enough, they get to own it. The artists are also automatically members of the cooperative. Decisions on what to do with the profits are taken democratically by all members. A third organizational model has emerged out of the cooperation with trade unions. Here, too, the best-known examples come from the USA. Several alternative taxi schemes have been set up that contrast with Uber; they include the Transunion Car Service, Union Taxi and the App-Based Drivers Association. Last, cooperative forms of organization can arise in partnership with cities and municipalities: examples include cooperative accommodation-finding platforms and intermodal apps for organizing and developing public local transport.

Policy-makers can foster the creation and expansion of platform cooperatives in a variety of ways, such as through tax concessions for start-ups and financial support schemes specifically geared to their needs. Cooperative undertakings can also be given preferential treatment in the awarding of public contracts. There is also scope for the state to sponsor research into successful platform cooperatives in order to identify success factors and thus promote upscaling. An even more ambitious strategy would involve pushing ahead with the "cooperative ecosystem"[19] that arises when cooperative platforms connect with each other and create synergies. This can be encouraged by policy-makers through networking events and the development of shared structures. Open-source applications are an example of this sort of sharing and are worth promoting because the availability of free hardware and software can be very useful when setting up cooperative platforms.[20] By promoting cooperative platform economies, policy-makers can make an important contribution to the guiding principle of a focus on the common good, since decision-making and profits are shared among a large number of beneficiaries.

Accelerate digitalization in selected areas

In addition to cooperative platform economies, there are other areas that policy-makers can promote to ensure that digitalization advances along lines that are as social-ecological as possible. These other areas include:

- applications and infrastructure for connected public and shared-use passenger transport (local public transport and sharing);
- applications and infrastructure for smart decentralized electricity grids (such as micro-grids) and sufficiency-oriented energy management systems for controlling heating systems;
- development of green apps that facilitate sustainable consumption;
- regional networking of producers and consumers in various sectors of the economy;
- development of open-source hardware and software.

Public subsidies in areas such as these can enable sustainable digital applications to achieve market readiness more quickly and thus increase their ability to compete with non-sustainable providers. Policy-makers can also help to improve data protection by promoting strategies based on privacy by design and data sufficiency. And, finally, policy-makers can establish and expand the relevant infrastructure for connected public transport, decentralized energy generation and so on.

A suitable promotional instrument is the awarding of public funds for research and development. Research grants relating to digitalization are currently awarded on a relatively non-specific basis and are rarely linked to social-ecological objectives.[21] Innovative developments can also be supported via incubator programmes or accelerator camps that help start-ups with funding and advice in the early stages. Policy-makers can also provide networking platforms to encourage dialogue and project development. A National Office for Social Innovation should be established in many countries on similar lines to the USA's Office of Social Innovation and Civic Participation to promote social-ecological sharing schemes and forms of prosuming.[22] Sustainable digitalization can also be encouraged through targeted public procurement. The public sector accounts for a large part of the demand for ICT devices. Public institutions thus have considerable power to shape the ICT market. Public contracts can specify that tenderers meet high standards relating to privacy, green IT and minimal energy consumption, thereby adhering to the guiding principles of strict data protection and digital sufficiency.

Aim to re-regionalize the economy

Local economic structures are often more environmentally sustainable than continental or global markets.[23] They also make it easier for consumers to gain insights into the conditions under which products are made, thus promoting more democratic management of the economy.[24] However, critics often say that local production is inefficient and, hence, too expensive. Digitalization provides new opportunities for organizing economic activities at the local level and solving the problem of lower economic efficiency. The prerequisite is that the increased labour productivity and greater resource and energy efficiency that digitalization makes possible are not translated into environmentally problematic economic growth, but are used to strengthen regional and local production. Instead of being used to boost mass production by multinational agribusinesses,[25] digital opportunities in agriculture could benefit local agro-ecological production – based around open-source farm machinery,[26] open-source seed[27] or digitally-supported local markets.[28] For example, GoCoop is an online social marketplace for India's cooperatives and community-based enterprises. It works with more than 70,000 producers, and aims to impact a million weavers and artisans working in rural areas.

By shaping the policy framework accordingly or using incentive mechanisms, policy-makers can, in many cases, influence whether digitalization serves to boost global economic efficiency or regionalization. Digital-ecological reform of the tax system (see next section) would do much to support local production. Multi-national companies build their businesses on tax avoidance and low transport costs. If companies like Amazon had to pay more tax and incurred higher transport costs, local providers would be better able to compete. The passivity rule and the advertising ban would also work in favour of local providers: compared with the large corporations, they rely less on internet advertising and more on word of mouth.

Municipalities can also play a part by using digital opportunities to improve connectivity in their localities. For example, as part of the German "Digital Villages" project, a mobile shop travels around small villages, with orders and routes being coordinated digitally.[29] Municipalities can also link all sorts of local service providers – from tradespeople and providers of social services to freelance software developers – with residents who need these services. The "Smart Countryside" study in Finland finds that digitalization can bring services nearer to the customer, reduce costs and have a major impact in the countryside, where structural change is rapid and distances to physical services are increasing.[30] Many services can then be brought together in a regional platform or app set up by the municipality: RegioApp and frimeo in Germany, regional.tirol in Austria and a community car-pooling initiative in Italy, which is managed by a local community cooperative,[31] are examples of schemes that are already in operation.

Re-regionalization further requires smart management and adaptation of traffic flows and infrastructure, because uncontrolled digitalization of transport is one of the structural factors leading to longer journeys and more travel in general. If we use digitalization, not only to regionalize economic processes, but also to support regional mobility, then the volume of traffic can be reduced while maintaining people's mobility. And, if this results in new local structures – such as shopping and leisure facilities – with living and working spatially integrated, we shall need to make fewer journeys, but will still be able to meet all our needs. This ultimately offers potential to slow down our pace of life and enhances life satisfaction.

Undertake a digital-ecological reform of the tax system

As well as taking specific steps to manage digitalization, it is essential to adapt the macroeconomic framework. As we have seen, digitalization is transforming many sectors of the economy. We will now show that a digital-ecological reform of the tax system, shorter full-time working hours and expansion of the care economy are key macroeconomic responses to these changes.

Thirty-five years ago, when personal computers were still a rarity, Swiss economist Hans Christoph Binswanger proposed a radical reform of the tax system that was intended to kill three birds with one stone. His concept of an "ecological tax reform" involves progressively increasing taxes on energy and resource use.[32] First, this gives businesses and consumers an incentive to use natural resources more sparingly. Second, the revenue can be used to reduce existing taxes on labour, such as pension contributions. This encourages businesses to create new jobs and reduce unemployment. Third, taxing resource use can broaden the tax base. The concept is now more relevant than ever. As we have shown, digitalization is built on the exploitation of scarce and rare resources and is accompanied by a dramatic increase in the demand for electricity. The virtual economy – like the "real" one – must therefore be given greater incentives to use energy and resources more sparingly. We have also shown that digitalization can lead to significant job losses. Making energy and resources more expensive relative to labour would increase the attraction of having work performed by humans rather than by machines and robots. In addition, we have seen that many global corporations behave like digital freeloaders and systematically avoid contributing to the general tax fund. A tax system that is fit for the future must address this issue.

We recommend that the concept of ecological tax reform be adapted to our increasingly digital economy and society and developed into a "digital-ecological tax reform". Instead of only taxing energy and resources, the main additional feature of this system would be that profits from digital value creation and automation would also be taxed. This is necessary, first, because the revenue from the ecotax might not be enough to reduce taxes on labour to the point where a sufficient incentive for new jobs is created. Second, it would broaden the revenue base and also remain reliable in the long term even if consumption of energy and resources starts to fall. Some representatives of the IT sector, such as Bill Gates, have proposed a "robot tax".[33] The former Greek finance minister, Yanis Varoufakis, along with economists such as Mariana Mazzucato, suggests that the corporations that build robots and other labour-displacing technologies benefit enormously from various forms of public investment and should therefore pay a "universal basic dividend" into a public trust. Effectively, society becomes a shareholder in every corporation and the resulting dividends could be distributed evenly to all citizens.[34] Brand profits could also be taxed: the main reason for the financial strength of Apple, for example, is that the company's image is so strong.

These are just a few ideas for a digital-ecological tax reform, the details of which would require careful thought. But, it is abundantly clear that the profits of digitalization must be taxed and – in accordance with the guiding principle of focusing on the common good – used in ways that benefit society as a whole. A digital-ecological tax reform can play an important part in steering digitalization in a green direction and distributing its profits more equitably.

Shorten full-time working

For most people in industrialized countries, gainful employment is the main source of livelihood security – and also a key source of personal satisfaction, social status and a feeling of belonging to society. However, full employment has for many years been an unrealistic goal. Although unemployment in several high-income countries is currently relatively low, there are still too many people out of work and many more who are underemployed. This situation will become more critical as a result of the likely rationalization effects of digitalization – even if all the scenarios that materialize merely involve relatively "moderate" net job losses (see Chapter 4). It must finally be recognized that a 40-hour-a-week job for everyone is an unachievable goal.

This challenge will only be addressed to a limited extent by a digital-ecological tax reform. In an increasingly digitalized society, policy-makers should therefore pursue a new labour policy that is based on principles of redistributive justice and enables both women and men to participate in gainful work within society. According to the International Labour Organization, a reduction of working hours could mean a shorter working day (e.g. a six-hour working day), a shorter working week (e.g. a four-day working week) and/or a shorter working year based on additional days of paid annual leave, paid holidays or other types of leave.[35] A redistribution of paid work along these lines improves the employment prospects of all those who have lost their jobs as a result of digitalization (or for other reasons) or are at risk of doing so. Of course, this also reduces the incomes of people who have until now worked a 40-hour week. In its place, they gain time: leisure time, family time or time to pursue other forms of work – perhaps in the care economy, which we shall discuss in the next section, or in the do-it-yourself economy. In any event, short full-time working will do much to establish solidarity as a basic principle underlying the organization of gainful employment within society. It counteracts the polarization of income and the precarization of large swathes of the population and thus makes an important contribution to the common good.

Short full-time working also improves the opportunities for equal gender-independent participation in all areas of work. This is partly because men are currently still more likely than women to be in full-time employment; many women work part-time. In addition, some career options and many top positions are still linked to full-time work and are therefore more often a male preserve. Short full-time working thus makes it easier for women to achieve equality in all areas of employment and at all career levels. At the same time, short full-time working for men would involve a reduction in working time and give them the opportunity to be more involved in caregiving (and remove an "excuse" for not doing more). All working people would have more time for non-remunerated pursuits: for volunteering, political engagement and cultural activities, and all forms of independent work – repairs and maintenance, self-build projects, upcycling, tending an allotment, urban gardening and so on. If this results in services that were previously organized via the market being undertaken instead on a subsistence or do-it-yourself basis, an important contribution to sufficiency will also have been made.

And, finally, a reduction in the time spent in paid employment provides an opportunity to slow the pace of life. This is another reason why enabling short full-time working in the digital society is more relevant than ever before. Reducing the time spent in paid employment broadens the scope for achieving a stress-free work-life balance and increasing overall life satisfaction. That way, the opportunities of digital efficiency and productivity improvements would not spur further economic growth, but a good life for all.

Expand the care economy

In the recent past, social status has come to be defined primarily by a person's job. In society at large, activities such as child rearing, care of the elderly and domestic work are seen as relatively low-status pursuits. This applies both to the caregiving that many people perform on a voluntary basis in looking after elderly or infirm family members and to the – often poorly paid – work in pre-schools, hospitals, care homes and nurseries. Most of this work – whether paid or unpaid – is still performed by women. This often imposes a double burden on them, since they must juggle the demands of their career and their caregiving activities.

Despite claims from the IT sector that artificial intelligence will soon lead to the development of "robot carers", social and personal services are bound to be an area in which the human touch will remain indispensable. In many countries, demand for these services will even increase in the coming decades as a result of demographic change. Both the introduction of short full-time working and the digital-ecological tax reform can help bring about a revaluation of caregiving. The revenue from the tax reform – and thus the productivity gains from digitalization – could (also) be used to provide financial and human resources for the care economy. This would address both the significant understaffing and the issue of low pay, thereby remedying the gender inequality in this extremely important area.

What can users do?

Whether and to what extent digitalization can make the world greener and more equitable depends, not only on the policy measures taken, but also on individual behaviour. Digital technologies undoubtedly open up countless opportunities to lead a more sustainable life. But, as economist John Maynard Keynes warned many years ago: "The difficulty lies not so much in developing new ideas as in escaping from old ones."[36] As citizens, how can we use digital applications in order to replace traditional environmentally polluting consumption and mobility habits with practices geared to sufficiency and the common good? What can we do ourselves to protect our data? Below, we suggest a few steps that each and every one of us can take.

Consume sustainable products with digital aids

As users, we have numerous opportunities to utilise the internet and a host of green websites and apps that support more sustainable consumption. Many have already been mentioned. For example, Oroeco helps you calculate your daily carbon footprint and the Energy Consumption Analyzer provides tips on saving electricity in the home. Food Made Good UK is a community platform working together to create a sustainable foodservice industry, including an online restaurant guide to find somewhere sustainable to eat. HappyCow recommends vegan and vegetarian food outlets. Mobility apps are springing up like mushrooms – for example, Via operates ride-sharing services in New York, Chicago and Washington, DC, London, Amsterdam and Berlin. A variety of companies offers rental bikes in cities around the world. All sorts of shopping advice sites, such as the Australian Goodfish and fair fashion guides like the Fair-FashionFinder, identify and rate sustainably produced consumer goods. Apps such as the Label App explain what the various sustainability labels mean, while Ecolabel Index is the largest online directory of ecolabels. The myENV app directs users in Singapore to the nearest collection point for electronic waste. And platforms such as utopia.de collate all this information and offer it from one source.[37]

In connection with data protection, users can exert influence by opting for providers that prioritize secure data handling and maximum privacy. Instead of using an email address with Gmail, where the contents of email traffic are analysed for advertising purposes, anyone can choose to use providers such as Riseup or Posteo that handle data carefully and, in addition, run their servers on electricity from renewable sources. Instead of "googling", we can use search engines such as Startpage or DuckDuckGo that do not analyse data for advertising purposes and do not deliver personalized results. Signal, Threema or other secure messaging apps are alternatives to WhatsApp.

Let's be clear: there is no longer any excuse. Making use of these sustainable consumption opportunities that are readily available via smartphone, tablet or PC requires no great effort. So, let's do it – and spread the word!

Use peer-to-peer sharing schemes

Digital applications create opportunities for people to work together – by lending, swapping or producing things jointly. A collaborative economy is based, not on relationships between companies and customers, but on relationships between individuals – "peer-to-peer". Swapping and sharing also enable products to be used for longer and/or more intensively.

The sharing economy has been the target of increasing criticism in recent years. This is not because sharing is in itself a bad idea, but because the concept has been appropriated by some "platform capitalism" providers whose business models have little in common with the give-and-take of peer-to-peer sharing. Platforms such as Airbnb and Uber, or commercial carsharing schemes provided by large automobile companies, all peddle the idea of sharing, but are, in fact, adopting new profit-oriented ways of doing business that have nothing to do with real sharing. Airbnb is becoming more and more of a Bed & Breakfast and hotel letting site; Uber is the new taxi company. Their business models contribute nothing to a shift in values that will encourage adoption of the sort of sufficiency behaviour that is one of the underlying principles of the sharing economy. Rather, they are a form of "sharing extractivism" that is based on exploitation: Uber makes substantial profits from arranging journeys while its drivers earn very little. Airbnb is now a high-value stock market-listed company, yet the homeowners do the actual work. These providers should not be posing as pioneers of the sharing economy. Quite the contrary, their operations conflict with the guiding principle of a focus on the common good.

However, there are alternatives. For example, Fairbnb is an evolving cooperative scheme to make short-term accommodation rentals more sustainable and community-driven. The platform is operated by a community of hosts and guests, local businesses and neighbours – although users cannot currently book accommodation directly via the site. BlaBlaCar in Europe, PopaRide in Canada or sRide in India are ride-sharing platforms that enable users to offer each other lifts, making them genuine examples of peer-to-peer sharing. Many other local and regional alternatives to Uber are now springing up.[38] As users, we can therefore make a distinction between "good sharing" and "bad sharing". The "good" sharing models serve peer-to-peer sharing and conform to the principles that we have described in connection with the common good and discussed in the section "Strengthen platform cooperatives".

Peer-to-peer sharing schemes make a genuine contribution to the guiding principle of digital sufficiency. Initial studies show that the environmental impacts of some peer-to-peer sharing models are significantly lower than those of products that are purchased new and not shared – even when the ecological footprints of digital devices are taken into account.[39] Clothes-swapping via platforms like Poshmark in the USA, Vinted in the UK or the Swancy App out of Norway has roughly half the eutrophication and greenhouse gas emissions associated with buying new items. Ride-sharing saves a significant amount of energy by comparison with unshared journeys in a private car – and, unlike Uber or conventional taxis, it does not generate any new journeys, but enables drivers to offer lifts on routes that they would be travelling anyway. In this respect, it is more like traditional hitch-hiking. But, again, with peer-to-peer sharing it is worth checking exactly what is involved. A study of film streaming that compares peer-to-peer models with commercial internet television concludes that sharing has a larger carbon footprint: the energy consumption of end-use devices per transmission unit is higher than that of servers, and in the case of file-sharing between users it is in fact twice as high.[40]

Generate social innovation

The term "social innovation" refers to new behaviours and consumption patterns that differ from current habits or routines and aim to support a social-ecological transformation.[41] Some sharing schemes are particularly strongly linked to changing social practices and, in some cases, would not have been possible at all without digitalization. In the tourism sector, couchsurfing is an example. Instead of booking a hotel or a holiday apartment, people – generally the young – can use the couchsurfing website as a straightforward means of arranging an overnight stay with strangers. This travel-related social innovation not only provides an alternative to a night in a hotel, which usually has a large environmental footprint, but leads to a completely different style of travelling – one which also provides new social contacts and unique insights into other cultures. Another example of a sharing model that is particularly socially innovative is homeswapping. Various platforms enable home-owners to swap apartments or houses on a private basis for holiday purposes.

The foodsharing app shows how digitalization makes social innovation possible in the food sector. It helps to stop supermarkets throwing away leftover food: the food is instead distributed to foodsharing members free of charge. The idea is based on something that has long been known. Post-harvest losses do not occur only on farms, at grain markets or in the wholesale trade: vast quantities of food are also thrown away in the final few metres of the supply chain – in shops or homes. Dumpster diving is legal in the USA, except where prohibited by local regulation. In other countries, some groups have been illegally engaged in "dumpster diving" for decades, retrieving discarded food from supermarket skips at night. When it comes to wasted food, the smartphone app Olio is at the top of its game, with 450,000 participants in 32 countries. Instead of local food shops and restaurants discarding products, they are picked up and posted on Olio.[42] Other great examples of food sharing apps are the apps La Piat, located in Canada, and Food Rescue, reducing food waste in the USA. The more than 200,000 registered users and 3,000 participating shops in Germany, Austria and Switzerland using the foodsharing app have been estimated to have prevented eight million kilogrammes of still-edible food from being thrown away.[43]

Another example relates to the organization of events. Long journeys – especially air travel – have a very large environmental footprint. The organizers of an international conference on resource management employed digital technology to significantly reduce the amount of travel undertaken by delegates. After lengthy debate about whether the conference should be held in Japan or in Austria, it was eventually held at both locations, which were linked by various forms of video technology so that participants could attend the same programme and talk to each other. Even after factoring in all the energy costs of the video technology and allowing for the fact that, with its two venues, the conference attracted more participants in total, CO_2 emissions were cut by an estimated

37–50 per cent.[44] As these examples show, digital technologies create scope for new forms of sustainable action. But, technology alone is not enough: transformation also requires creativity and a willingness to embrace change on the part of many individuals and groups.

Slow the pace of life

The use of digital media in more and more areas of our private and working lives is speeding up our pace of life. Initially, we save time by using digital technology: we can communicate more quickly, find the best route to our destination, shop online with a click of a mouse and so on. But, in the end, rationalizing our use of time can generate a time-use rebound effect: our lives are constantly becoming faster and more complex because we are performing more and more private and work-related tasks in ever-shorter periods of time or even simultaneously. We, the authors, are convinced that "more of the same" is not the right formula for addressing the systemic impacts of time-saving technology. Using even more apps and digital assistants so that we can shop, travel or communicate even faster will simply worsen the vicious circle. The answer lies instead in intelligent self-restraint. Only by exercising moderation in our use of technology, as well as in our everyday planning, can we curb the dynamics of digital overload and make space for new ways of doing things. The first step towards slowing our pace of life involves having the courage to spend more time offline – in other words, to do a "digital detox".

For all of us who want to carry on using our smartphones and computers to manage our work and daily lives, this means that, the more we use time-saving digital applications at home or at work, and the more adept we become at doing so, the longer we need to keep them switched off if we really want to create space for leisure, social interaction and down-time. Aren't these the very things that time-saving technology was supposed to give us? Wasn't it intended to save us time at work and time spent on errands or organizing meetings so that we could then work less, spend less time on chores and have more time for friends, family and ourselves? The technology will not deliver these rewards by itself – we must take responsibility. We must transform our user behaviour and start to practise balance and moderation in our use of digital tools.

Nevertheless, it will sometimes be difficult to break out of this cycle of ever-faster activity. After all, we are not only speeding up our lives ourselves; everything around us is also constantly urging us to up the pace. We, therefore, need a corporate culture that does not look for increased output when we use our time efficiently at work, but, instead, allows us to go home earlier. The new opportunities for work that is flexible in terms of time and place should not result in demands for ever-greater blurring of the boundaries between our work and private lives, but should, instead, allow us clearly demarcated work-free periods. Businesses can encourage this by blocking the delivery of emails after a particular time in the evening and at weekends. Employees who finish work earlier in the day because they have worked efficiently then have a good chance of actually benefiting from the time saved. A policy framework designed to both reduce working hours and clearly separate work from leisure time is a key driver of a slower pace of life in the digital society.

And, who knows: outside the world of work, perhaps we shall one day hear calls for a "digital detox" in public places. The ban on smoking in planes, at railway stations and in restaurants was for many years decried even by non-smokers as a restriction of personal freedom. But, nowadays, almost everyone recognizes that precisely the reverse is true and that bans on smoking in public places extend personal freedom. This idea can be transferred to digitalization. Currently, the trend is still towards making Wi-Fi and broadband available in every nook and cranny of our lives. Laptop-free cafés that draw people in by advertising themselves as "digital-free" spaces are pioneers in the move towards moderation in digitalization.

Civil society is key

Policy instruments and an enabling regulatory environment are the most influential means to guide economic and social developments. Sustainability-oriented behaviour on the part of users is also essential if we are to transform digitalization. However, both policy-makers and users often need a push, or indeed a long series of pushes, in order to change course. This is one of the classic tasks of civil society. We believe that associations, NGOs, movements and civil society networks have a key part to play in the sustainable transformation of society and could be more actively involved in steering digitalization in a social-ecological direction.

Shape debates on digitalization

It is true that there is already very lively civil society engagement on digital and internet issues the world over. This starts by an array of user groups (e.g. Linux user groups), through many diverse hacker groups, to non-governmental organizations focusing on network neutrality, surveillance, data retention, the Snowden revelations, copyright, open source, hate speech and other issues. Many of these groups are not afraid to voice criticism and use their accumulated expertise to offer proposals for public discussion. However, as we have seen, digitalization affects many other issues relating to the future of our planet – particularly the environment, but also work, inequality, gender equality and global justice – that are not (yet) at the top of their political agenda.

Considering the scale of possible developments, many more of the civil society organizations that have considerable expertise in these areas should become involved in the debate about the future of the global digital society. A number of activities and publications, driven partly by trade unions, have addressed the impacts on the labour market. With regard to other issues, though, progress has been slow, although some initial steps are being taken. For example, Green-peace has breathed new life into the longstanding debate on green IT with its reports on the environmental impacts of the internet[45] and smartphones.[46] Trade union organizations lobby on digital work issues. The Freelancers Union offers unique services for freelancers in the USA. The London-based Couriers and Logistics Branch of the Independent Workers of Great Britain is doing pioneering work defending the rights of workers in the British courier and logistics industry. Unionen (Sweden) has developed a plan to certify platforms for fair and socially sustainable working conditions. The Austrian ÖGB and The Austrian Chamber of Labor are doing pioneering work supporting crowd-workers,[47] and some also have begun to tackle the issue of digitalization and gender equality.[48] The list of individual activities could be continued, but the inescapable conclusion is that, given the rapid pace of digital developments and their potentially radical social and environmental impacts, civil society needs to play a far more active role.

In particular, there is a lack of strong civil society voices focusing on the environmental aspects of digitalization. The question of what the digital world society will look like in future and how this could change the metabolism of our industrialized society should not be left solely to the business giants of Sili-con Valley, the venture capitalists and start-ups. Environmental and development organizations, churches, trade unions, welfare organizations, social movements and think tanks working in the applied sustainability sciences can contribute to the discussion in two ways, as they do in other contexts. First, they can act as critical companions: they can do far more to anchor, not only the opportunities, but also the risks of digital developments in the public debate, ensuring that these issues can no longer be ignored by policy-makers and users. Second, they can be agenda-setters: where the debate has focused one-sidedly on the promise of growth, they can highlight all the social and environmental challenges that we have mentioned. The trade unions have already succeeded in bringing the ques-tion of digitalization and jobs to the fore. But, other issues are often ignored and this narrows the focus, with the result that policy-makers have yet to undertake any comprehensive assessment or governance of digital developments. A critical civil society can step in here, identify alternatives and thus persuade people that alternative digitalization is possible.

Promote critical digital education

The reflex reaction to the challenges of digitalization generally involves calls for more education: schoolchildren should learn about technology and the inter-net, vocational training should be expanded to include digital content, people should be re-trained for IT jobs and digital courses should be available for older people as well. However, the call for digital education is often based, not on Humboldtian ideals, but on economic interests. In view of the rising demand for programming and other IT-related expertise, the expansion of digital education will, it is claimed, help alleviate the potential shortage of skilled workers in these areas.[49]

However, critical digital education should have other goals in mind. One that has already received frequent mention is that of "media competence"; this involves educating people in the use of digital products and services in order to prevent learning and attention deficits or forms of internet addiction, for example.[50] Digital education also involves helping consumers to acquire "digital sovereignty", so that they are aware of privacy issues, are careful about what they do with their data and can assert their rights relating to online contracts.[51] But, critical digital education can go further. It should highlight the opportunities for social-ecological change and for a good life that digitalization can unlock – from sustainable online consumption via local product providers and the use of alternative, cooperative platforms to ways of becoming a prosumer oneself. Civil society organizations can help people make decisions. They can publish consumer advice that explains which applications misuse data or are particularly energy-intensive. They can organize events to discuss alternatives to capitalist platforms and ways of using the internet without compromising personal privacy. And they can run campaigns that provide information on the amount of energy and resources consumed by the various devices and applications, highlight ways of reducing this consumption and signpost open workshops and open-source information so that people can learn how to extend the useful life of their devices themselves.

At the same time, critical digital education should ask questions about our society as a whole. What sort of digitalization do we want, and how much of it should there be? Who wins and who loses in the digital revolution? What does the digitalization of consumption and industry mean for the environment? What impacts does it have on the global South? Ideally, these questions should also be addressed in state schools and colleges. But, past experience has shown us that genuine critical debate requires the involvement of an active civil society. Furthermore, the developments associated with digitalization are happening so fast that the state education system can barely keep up.

Critical education faces a particular challenge in connection with digitalization, because memorable images are harder to come by here than in other areas. What is the equivalent of polar bears or whales, nuclear reactors, cleared forests or – turning to positive images – the view of our fragile Blue Planet from space? Discussion of the impacts of digitalization is beset by the problem of "psychological distance" already familiar to us in the context of climate change. If we get in our car and drive to the bookshop to buy a book, we are aware that it uses petrol. But, if we have the book delivered, the direct impacts are outside the field of our personal perception and need to be processed at an additional level – that of reflection – before we change our behaviour. We also lose sight of the social costs. When we deal with the saleswoman in the shoe shop, we may perhaps get a glimpse of her working conditions; we have no such information about the workers in the Zalando warehouse. The same applies to all the other goods and services that have been shifted from the physical to the virtual space. The increasing psychological distance that accompanies the digitalization of consumption makes it more difficult to deliver critical education and engage users. A successful civil society will encourage people to take an interest in these hidden effects of digitalization and take them seriously as consequences of their consumer decisions.

Form a broad movement

For social change to be successful, organizations that engage at the discursive and political level need to collaborate with those that develop practical solutions. Showing how things could be done differently right here, right now helps to drive development. This has been the case with the transition to sustainable energy. Yes, environmental organizations and movements have played an extremely important role here, but without all the DIY enthusiasts and inventors who built wind turbines and solar energy systems in garages and backyards and tested them single-handedly, and without the small progressive engineering firms, businesses and "electricity rebels" that set out to help mitigate climate change and contribute to the common good, the transition would not have progressed so rapidly. Nothing is more convincing than seeing the solar energy system or the wind turbine right there in front of us and realising that it can be done!

Likewise, people who are developing theories and strategies for an alternative economic order are now cooperating with initiatives, companies and projects that are already adopting environmentally sustainable, solidarity-based or non-commercial approaches. This helps to link discussion of the post-growth economy, the solidarity-based economy, the commons, transition towns, the economy for the common good and peer-to-peer sharing with practical projects in the fields of solidarity agriculture, give-away shops, cooperatives, collective businesses and barter communities. Some are working to raise social awareness, while others offer practical examples of how things can be done differently.

In connection with digitalization, too, theorists and practitioners can do more to work together for mutual benefit. We have already mentioned some of the players that are seeking to influence policy-making and are engaged in public debate: NGOs, churches, trade unions, welfare organizations, social movements and other groups working on internet-related issues and the social, environmental or gender equality impacts of digitalization. On the other side, there are countless projects, initiatives and alternatively-minded businesses that are developing practical applications, organizing projects or getting cooperative start-ups off the ground and, in so doing, modelling a different approach to digitalization. There is, for instance, the open-source community, already mentioned, that has been working on software for decades and is now turning its attention to hardware as well. Closely associated with this is the maker movement, whose members (re)learn in "fablabs" and open workshops how to make and repair things themselves.[52] There are also countless socially and environmentally motivated developers of apps and business ideas in the digital space. They would all be obvious partners for a civil society movement working to change the direction of digitalization.

When different "scenes" and "subcultures" start working together, this often unleashes particularly innovative and creative synergies that can trigger, not only technical progress, but also a move towards a freer society. There are thousands of civil society initiatives, organizations and networks that aim to use digitalization, not for surveillance or commerce, but for a better world. If they pool their energies, this will create a powerful force that even the world's richest IT companies cannot ignore.

Notes

1 European Commission, The Digital Agenda for Europe, 2012.
2 Acker, Data Craft, 2018; Ross *et al.* Are social bots a real threat?, 2019.
3 Rubin *et al.*, Covert Communication in Mobile Applications, 2015.
4 European Commission, *Regulation of the European Parliament and of the Council on the protection of natural persons with regard to the processing of personal data and on the free movement of such data (General Data Protection Regulation)*, 2016.
5 European Commission, *Regulation of the European Parliament and of the Council on the protection of natural persons with regard to the processing of personal data and on the free movement of such data (General Data Protection Regulation)*, 2016.
6 Federal Trade Commission, Understanding mobile apps, 2017; Pew Research Center, Apps Permissions in the Google Play Store, 2015.
7 Gibbs, WhatsApp sharing user data with Facebook would be illegal, rules ICO, 2018.
8 European Union Agency For Network and Information Security, Privacy and data protection in mobile applications, 2017.
9 EDRi, ePrivacy: EU Member States push crucial reform on privacy norms close to a dead end, 2019.
10 Privacy Rights Clearinghouse, Smartphone Privacy, 2005/2017.
11 Sachverständigenrat für Verbraucherfragen, Verbraucherrecht 2.0 Verbraucher in der digitalen Welt, 2016, p. 67.
12 Jarrahi/Sutherland, Algorithmic Management and Algorithmic Competencies, 2018.
13 Advisory Council for Consumer Affairs, Digital Sovereignty, 2017.
14 See e.g. *The Economist*, Regulating the internet giants, 2017.
15 E.g. Jewell, Digital pioneer, Jaron Lanier, on the dangers of "free" online culture, 2016; Orlowski, Jaron Lanier, 2016; the *Guardian*, Tech giants may have to be broken up, says Tim Berners-Lee, 2018.
16 Mazzucato, Let's make private data into a public good, 2018.
17 Scholz, Platform Cooperativism, 2016a.
18 Saner/Yiu/Nguyen, Platform Cooperatives, 2018.
19 Scholz, Platform Cooperativism, 2016a.
20 Scholz, Platform Cooperativism, 2016a.
21 Holzer *et. al*, Negotiating local versus global needs in the International Long Term Ecological Research Network's socio-ecological research agenda, 2018.

22 World Economic Forum, Breaking the Binary, 2013.
23 See e.g. in Sachs/Santarius, *Slow Trade*, 2007.
24 Manzini, Resilient systems and cosmopolitan localism, 2013, p. 77.
25 See on this Mooney/ETC Group, Blocking the chain, 2018.
26 Open Source Ecology, Open Source Ecology, 2017.
27 OpenSourceSeeds, OpenSourceSeeds, 2017.
28 See e.g. Pesce *et al.*, Research for AGRI Committee, 2019.
29 European Network for Rural Development, Digital Villages Germany, 2018a.
30 European Network for Rural Development, Smart Countryside study Finland, 2018b.
31 European Network for Rural Development, Strategy for Inner Areas Italy, 2018c.
32 Binswanger *et al.*, *Arbeit ohne Umweltzerstörung*, 1983.
33 See e.g. Kharpal, Bill Gates wants to tax robots, but the EU says, 'no way, no way', 2017.
34 See e.g. Dunlop, What is a robot exactly – and how do we make it pay tax?, 2017.
35 Messenger, Working time and the future of work, 2018.
36 This quote is widely attributed to John Maynard Keynes, e.g. in: Keynes, The General Theory of Employment, Interest, and Money, 1936/2003.
37 A detailed list of such applications is provided by Brauer *et al.*, Green By App, 2016.
38 Carey, Alternatives to Uber: The best alternative ride-hailing apps, 2017; Solon, Is Lyft really the "woke" alternative to Uber?, 2017.
39 Ludmann, Ökologie des Teilens. Bilanzierung der Umweltwirkungen des Peer-to-Peer Sharing, 2018.
40 Hochschorner/Dán/Moberg, Carbon footprint of movie distribution via the internet, 2015.
41 Social Innovation Community, What is Social Innovation?, 2018; Stanford Graduate School of Business, Defining Social Innovation, 2019.
42 Lopez, The 3 Best Food Sharing Apps, 2019.
43 Höfner/Santarius, Wertschätzungs- statt Wegwerfgesellschaft, 2017.
44 Coroama/Hilty/Birtel, Effects of Internet-based multiple-site conferences on greenhouse gas emissions, 2012.
45 Greenpeace, Clicking clean, 2014; Greenpeace, Clicking Green, 2017a.
46 Greenpeace, From Smart To Senseless, 2017b.
47 Fair Crowd Work, Shedding light on the real work of crowd-, platform-, and app-based work, 2017.
48 Ahlers *et al.*, Genderaspekte der Digitalisierung der Arbeitswelt, 2017.
49 Federal Ministry for Economic Affairs and Energy, Securing of Skilled Labour, 2019; McKinsey Global Institute, Digital India: Technology to transform a connected nation, 2019.
50 Gutiérrez/Tyner, Media Education, Media Literacy and Digital Competence, 2012.
51 Gueham, Digital Sovereignty, 2017.
52 Smith *et al.*, Grassroots Innovation Movements, 2016, p. 109; Kohtala, Making "Making" Critical, 2017.

7
TOWARDS SOFT DIGITALIZATION

Digital technologies are changing our world, our workplace, our social relations and our economic structures with full force and at great speed. The resulting challenges add to the existing list of fundamental problems that will anyway dominate the agenda over the coming decades: crises, wars and regional instability are increasing worldwide, driving millions of people out of their homes and into a precarious future. Climate change, species extinction, degradation of fertile soil and urban air pollution are stacking up a cascade of environmental problems that will be our legacy to our children and grandchildren. And the increasing polarization of society into the haves and the have-nots – those with assets, on the one side, and those who fear for their jobs, income security and their rightful place in society, on the other – tends to undermine the principle of social solidarity. No matter of the specific role of digital apps and technologies: all these challenges call for fundamental changes to the way we run our economy, how we consume, how we share the benefits of prosperity and what we do about social welfare.

We started this book with a question: can digitalization help in overcoming the major challenges of our time? Now that we have reached its conclusion, let's take a step back and ask another question: what kind of social and environmental contribution can we expect digitalization to make? Can it help to put society on a transformative path towards sustainability? There is no straightforward answer to that question – that much has become clear. The individual technologies and applications are too disparate and their effects too ambivalent. So, our conclusion is more modest: the mega-trend of digitalization – particularly in the form it has taken over the last five to ten years – will not, in and of itself, solve any of the major challenges facing society. On the contrary, although digitalization does indeed create new opportunities, there is a risk that, under current economic and political conditions, it will worsen some of the problems societies are facing, such as income disparity, labour market insecurity, the risks of surveillance and intimidation, and growing consumption of scarce resources and climate-damaging fuels. Unless policy-makers, business makers, civil society and users intervene with real purpose, digitalization is likely to intensify these problems.

This does not mean that people should stop using smartphones or the internet. Nor would a blanket decision to refrain from using robots or artificial intelligence be any help. Instead, the best approach is to give much more careful and focused thought to the question of which digital applications help society move forward and which of them – notwithstanding promises that they will help to build a better future – are of dubious benefit. We should envision digitalization as a large and complex toolbox: some of the tools can solve some of the problems, but there is no single tool to solve every problem. Despite appearances, sometimes the tools don't fit and under no circumstances should the availability of tools determine what is defined as a social problem. On the contrary, it is necessary to clearly define desired outcomes first and, then, develop toolkits that offer the best solutions. With that aim in mind, digitalization must be managed much more intensively, much more selectively and much more critically by policy-makers and society than at present. The questions "What kind of digitalization do we want?" and "What do we want to use digital tools for?" should inform every discussion around this topic.

The frameworks put in place by policy-makers have a key role to play in this context. Several of the solutions that we have identified in this book have been under discussion for some time, but, in light of the changes brought about by digitalization, we believe they are more relevant than ever. For example, digital-ecological tax reform harks back to the more familiar ecological tax reform. The topic of shorter full-time working hours was under discussion some years ago – it simply needs to be updated and adapted to our digital society. These overlaps show that digitalization is not a single issue that puts all the old ones in the shade. On the contrary, it adds a new dimension to existing issues. So, the key question is not how we deal with digitalization in isolation, but how we can adapt existing ideas about social transformation to take account of digitalization and its impacts.

And another insight: the rapid pace of digitalization that many are predicting may be part of the problem, not part of the solution. At the outset of this book, we asked whether the disruptive potential of digitalization could facilitate a rapid and deep transition towards sustainability. But, although we touched on this aspect in our introduction, our readers will have searched in vain for ideas on how digital "disruption" may benefit the environment and social justice. That is because many of the developments being driven forward at astonishing speed, and with considerable force and financial input, do not take us in a sustainable direction at all. What is certain, however, is that many of the cooperative privacy- and sustainability-oriented solutions require tireless support to become established. Users, purpose-driven start-ups, policy-makers and civil society need time to make use of digital opportunities and to develop digital governance options. In this respect, they are clearly lagging far behind the IT companies, which are moving forward at lightning speed.

In short, there is a genuine risk that digital disruption will have too many negative effects on society to have any real value in the sense of increasing our rights and freedoms. Similar dynamics have been observed with other megatrends, notably globalization. After the Second World War, free world trade was initially viewed as a means of bringing about and promoting peace, opening up economic opportunities and improving intercultural understanding. But, since the Iron Curtain came down, free trade has become the domain of transnational corporations, which oftentimes maximize profits across borders while exploiting people and the planet. And what is happening now? Many of the people who voted for Trump in the USA or Brexit in the UK – along with millions of people in other countries, North and South – are feeling massive frustration. They see themselves – rightly or wrongly – as the losers of globalization. This opens the way for populist parties with xenophobic and anti-democratic agendas to enter our parliaments; and it also divides societies. And, yet, trade and economic globalization between countries is no bad thing as such. It is simply that society and politics has missed the opportunity to stop it becoming too much of a good thing. A similar trend is now emerging with digitalization. Disruptive and overly capitalist digitalization may well mean that even more people are left behind and lose their place in society. We must try to stop that happening.

In a recent interview, Astro Teller, Chief Strategy Officer at Google X, was asked whether society was still capable of keeping up with the rapid developments taking place in the technological field. He answered:

> Just recently, the rate of radical change in technology has been faster than the rate that society can think about it. But rather than seeing the technology as the problem, I would propose that what we actually need to do is to strengthen our ability as society, to think faster, to evolve more quickly, to adapt to the technical world; because I think that is more productive than slowing the technology down.[1]

This is a fairly typical view for many champions of Silicon Valley and the corporate world of IT. It is one that tends to be put forward by well-educated androcentric technophiles, who benefit from this approach. What they seem to ignore, however, is that traits such as empathy, but also social diversity and the speed of biochemical regeneration cycles, simply do not lend themselves to rapid or radical change. And, if we assume that Teller's is a minority view and that the majority of people believe in values such as democratic decision-making, the common good and sustainability, it is clear that we need to adapt digitalization to our notions of what constitutes a healthy society, not vice versa!

So, our plea is for soft, not disruptive, digitalization. Only soft and mindful digitalization that is clearly designed to contribute sustainable solutions to social challenges and helps to satisfy everyone's needs, regardless of background, education and income, will take the strain off the environment, inspire people with courage and strengthen social cohesion. We do not need a society of zeros and ones. We need a digitalization at a human scale.

Note

1 Kleber/Andersen, ZDF-Reportage: Silicon Valley, 2016; Interview from approx. minute 8:00 onwards.

BIBLIOGRAPHY

Acemoglu, Daron and Pascual Restrepo (2016): *The race between machine and man: Implications of technology for growth, factor shares and employment.* Cambridge, Mass: National Bureau of Economic Research Working Paper 16–05.

Acemoglu, Daron and Pascual Restrepo (2017): *Robots and Jobs: Evidence from US labor markets.* MIT Department of Economics Working Paper 17–04. Boston: MIT, Boston University.

Achachlouei, Mohammad Ahmadi and Åsa Moberg (2015): Life Cycle Assessment of a Magazine, Part II: A Comparison of Print and Tablet Editions. *Journal of Industrial Ecology* 19, No. 4 (1 August): 590–606.

Acker, Amelia (2018): Data Craft: The Manipulation of Social Media Metadata. Data & Society. https://datasociety.net/wp-content/uploads/2018/11/DS_Data_Craft_Manipulation_of_Social_Media_Metadata.pdf (accessed: 22 November 2019).

Adams, A. and Berg, J. (2017): When home affects pay: An analysis of the gender pay gap among crowdworkers. *SSRN Electronic Journal.*

Advisory Council for Consumer Affairs (2017): Digital Sovereignty. Available at: www.svr-verbraucherfragen.de/en/wp-content/uploads/sites/2/Report.pdf (accessed: 23 November 2019).

Aguiar, Mark and Erik Hurst (2007): Measuring Trends in Leisure: The Allocation of Time Over Five Decades. *The Quarterly Journal of Economics* 122, Nr. 3 (1 August): 969–1006.

Ahlers, Elke, Christina Klenner, Yvonne Lott, Manuela Maschke, Annekathrin Müller, Christina Schilmann, Dorothea Voss and Anja Weusthoff (2017): Genderaspekte der Digitalisierung der Arbeitswelt. Diskussionspapier für die Kommission "Arbeit der Zukunft". Berlin: Hans-Böckler-Stiftung.

Ahmadi Achachlouei, Mohammad, Åsa Moberg and Elisabeth Hochschorner (2015): Life Cycle Assessment of a Magazine, Part I: Tablet Edition in Emerging and Mature States. *Journal of Industrial Ecology* 19, No. 4 (1 August): 575–589.

Akos, Kokai (2014): *Whole Earth Catalog (1975)* Available at: https://s3.amazonaws.com/green buildingadvisor.s3.tauntoncloud.com/app/uploads/2018/08/07230306/15604607615_da25b7e97d_k_0-main-700x525.jpg (accessed: 6 December 2019).

Alcott, Blake (2008): The sufficiency strategy: Would rich-world frugality lower environ-mental impact? *Ecological Economics* 64, Nr. 4 (February): 770–786.

Alstone, Peter, Dimitry Gershenson and Daniel M. Kammen (2015): Decentralized energy systems for clean electricity access. *Nature Climate Change* 5, Nr. 4 (April): 305–314.

Altieri, Miguel (1995): *Agroecology: The science of sustainable agriculture*. Boulder: Westview Press.

Amadeo, Kimberly (2019): Income Per Capita, With Calculations, Statistics, and Trends. *The Balance*. Available at: www.thebalance.com/income-per-capita-calculation-and-u-s-statistics-3305852 (accessed: 28 October 2019).

Amatuni, Levon, Juudit Ottelin, Bernhard Steubing and José Mogollon (2019): Does car sharing reduce greenhouse gas emissions? Life cycle assessment of the modal shift and lifetime shift rebound effects. *arXiv:1910.11570 [econ, q-fin]* (accessed 25 October 2019).

Anders, Melissa (2018): Retail Industry Expects More Sales Growth In 2018. *Forbes*. Available at: www.forbes.com/sites/melissaanders/2018/02/08/retail-industry-poised-for-more-sales-grow9th-in-2018/ (accessed: 4 November 2019).

Andrae, Anders S. G. and Tomas Edler (2015): On Global Electricity Usage of Communi-cation Technology: Trends to 2030. *Challenges* 6, Nr. 1 (30 April): 117–157.

Apple (2009): iPhone 3G Environmental Report.

Apple (2015): Environmental Report iPod Touch (6th Generation).

Apple (2016): Annual Report.

Apple (2017): iPhone 7 Environmental Report.

Apple (2019): Product Environmental Report iPhone 11. Available at: www.apple.com/euro/environment/pdf/a/generic/products/iphone/iPhone_11_PER_sept2019.pdf (accessed: 4 November 2019).

Asdecker, Björn (2015): Returning mail-order goods: analyzing the relationship between the rate of returns and the associated costs. *Logistics Research* 8, No. 1 (December).

Aslam, Salman (2017): Instagram by the Numbers: Stats, Demographics & Fun Facts. *omni-coreagency.com*. 21 June. Available at: www.omnicoreagency.com/instagram-statistics/ (accessed: 3 December 2019).

Auerbach, Marc (2016): IKEA: Flat pack tax avoidance. *Morgunblaðið*. Available at: www.mbl.is/media/28/9628.pdf (accessed: 4 November 2019).

Autor, David, David Dorn, Lawrence Katz, Christina Patterson and John Van Reenen (2017): Concentrating on the Fall of the Labor Share. Cambridge, MA: National Bureau of Economic Research.

Balde, C. P., R. Kuehr, K. Blumenthal, S. Fondeur Gill, M. Kern, P. Micheli, E. Magpantay and J. Huisman (2015): E-waste statistics: Guidelines on classifications, reporting and indicators. United Nations University, IAS-SCYCLE, Bonn, Germany.

Barcham, Raphael (2014): Climate and Energy Impacts of Automated Vehicles. Sacramento: California Air Resources Board.

Bartmann, Christoph (2016): *The Return of the Servant*. Munich: Carl Hanser Verlag. Avail-able at: www.hanser-literaturverlage.de/en/buch/the-return-of-the-servant/978-3-446-25287-5 (accessed: 6 December 2019).

BBC (2017): Ukraine power cut 'was cyber-attack'. www.bbc.com/news/technology-38573074 (accessed: 4 November 2019).

Becker, William J., Belkin, Liuba, Tuskey, Sarah (2018): Killing me softly: Electronic com-munications monitoring and employee and spouse well-being. *Academy of Management Proceedings*. Vol. 2018, No. 1.

Beier, Grisha, Silke Niehoff, Tilla Ziems and Bing Xue (2017): Sustainability Aspects of a Digitalized Industry. *International Journal for Precision Engineering and Manufacturing-Green Technology* 4, No. 2: 227–234.

Bentham, Jeremy, and Miran Božovič (1995): *The panopticon writings*. New York: Verso.

Bentolila, Samuel and Gilles Saint-Paul (2003): Explaining movements in the labor share. *Contributions in Macroeconomics* 3, No. 1.

Bentzen, Jan (2004): Estimating the rebound effect in US manufacturing energy consumption. *Energy economics* 26, No. 1: 123–134.

Berg, J. (2015): Income security in the on-demand economy: Findings and policy lessons from a survey of crowdworkers. *Comparative Labor Law and Policy Journal*, 37, 543.

Berners-Lee, Tim (2016): The past, present and future. *Wired UK*. Available at: www.wired.co.uk/article/tim-berners-lee (accessed: 3 December 2019).

Bihr, Peter (2017): *View Source Shenzhen*. Berlin: The Waving Cat GmbH.

Binswanger, Hans Christoph, Heinz Frisch, Hans G. Nutzinger, Bertram Schefold, Gerhard Scherhorn, Udo E. Simonis and Burkhard Strümpel, Eds (1983): *Arbeit ohne Umweltzerstörung. Strategien für eine neue Wirtschaftspolitik*. Frankfurt: Fischer.

Binswanger, Mathias (2001): Technological progress and sustainable development: what about the rebound effect? *Ecological economics* 36, No. 1: 119–132.

Bischoff, Joschka and Michal Maciejewski (2016): Simulation of City-wide Replacement of Private Cars with Autonomous Taxis in Berlin. *Procedia Computer Science* 83: 237–244.

Blanco, Herib and André Faaij (2018): A review at the role of storage in energy systems with a focus on Power to Gas and long-term storage. *Renewable and Sustainable Energy Reviews* 81 (January): 1049–1086.

Bögeholz, Harald (2017): Künstliche Intelligenz: AlphaGo Zero übertrumpft AlphaGo ohne menschliches Vorwissen. *heise online*. 19 October. Available at: www.heise.de/newsticker/meldung/Kuenstliche-Intelligenz-AlphaGo-Zero-uebertrumpft-AlphaGo-ohne-menschliches-Vorwissen-3865120.html (accessed: 10 December 2019).

Bollier, David (2014): Will Bitcoin and Other Insurgent Currencies Reinvent Commerce? Available at: www.bollier.org/blog/will-bitcoin-and-other-insurgent-currencies-reinvent-commerce (accessed: 3 December 2019).

Bollier, David and Silke Helfrich (2019): *Free, fair, and alive: the insurgent power of the commons*. Gabriola Island: New Society Publishers.

Borenstein, Severin (2013): A microeconomic framework for evaluating energy efficiency rebound and some implications. *E2e Working Paper Series*.

Brand, Ulrich and Markus Wissen (2012): Global Environmental Politics and the Imperial Mode of Living: Articulations of State-Capital Relations in the Multiple Crisis. *Globalizations* 9, No. 4 (1 August): 547–560.

Brand, Ulrich and Markus Wissen (2018): *The limits to capitalist nature: theorizing and overcoming the imperial mode of living*. Transforming capitalism. London: Rowman & Littlefield International.

Brauer, Benjamin, Carolin Ebermann, Björn Hildebrandt, Gerrit Remané and Lutz M. Kolbe (2016): Green By App: The Contribution of Mobile Applications To Environmental Sustainability. *ECIS 2016 Proceedings*: 1–16.

Brown, Jay R. and Alfred L. Guiffrida (2014): Carbon emissions comparison of last mile delivery versus customer pickup. *International Journal of Logistics Research and Applications* 17, No. 6 (2 November): 503–521.

Brynjolfsson, Erik and Andrew McAfee (2012): *Race Against the Machine: How the Digital Revolution is Accelerating Innovation, Driving Productivity, and Irreversibly Transforming Employment and the Economy*. Lexington, Massachusetts: Digital Frontier Press.

Brynjolfsson, Erik and Andrew McAfee (2014): *The Second Machine Age: Wie die nächste digitale Revolution unser aller Leben verändern wird*. Kulmbach: Plassen-Verlag.

Bull, Justin G. and Robert A. Kozak (2014): Comparative life cycle assessments: The case of paper and digital media. *Environmental Impact Assessment Review* 45 (February): 10–18.

Bundesinstitut für Bau-, Stadt- und Raumforschung (2017): Online-Handel – Mögliche räumliche Auswirkungen auf Innenstädte, Stadtteil- und Ortszentren. BBSR-Online-Publikation 08/2017. Bonn: Bundesamt für Bauwesen und Raumordnung.

Bundesministerium für Verkehr und digitale Infrastruktur (2017a): Ethics Commission. Automated and Connected Driving. Available at: www.bmvi.de/SharedDocs/EN/pub lications/report-ethics-commission-automated-and-connected-driving.pdf (accessed: 11 November 2019).

Bundesministerium für Verkehr und digitale Infrastruktur (2017b): "Eigentumsordnung" für Mobilitätsdaten? Eine Studie aus technischer, ökonomischer und rechtlicher Perspektive. Berlin.

Bundesministerium für Verkehr und digitale Infrastruktur (2017c): Freight Transport and Logistics Action Plan – Towards a sustainable and efficient future, 3rd update 2017. Berlin: BMVI.

Bundesministerium für Wirtschaft und Energie (2019): Industrie 4.0. Digitally driven and smartly networked. www.bmwi.de/Redaktion/EN/Dossier/modern-industry-policy. html (accessed: 1 November 2019).

Bundesministerium für Wirtschaft und Technologie (2010): Energiekonzept für eine umweltschonende, zuverlässige und bezahlbare Energieversorgung. Berlin.

Bundesregierung (2016): *Gesetz zur Digitalisierung der Energiewende.*

Bundesverband der deutschen Versandbuchhändler (2015): Infografik: Amazon dominiert den Onlinebuchhandel. *Statista Infografiken.* Available at: https://de.statista.com/ infografik/2271/umsatz-des-versandhandels-mit-buechern-in-deutschland/ (accessed: 3 December 2019).

Bureau of Labor Statistics (2018a): Occupational Employment Statistics. May 2018. Available at: www.bls.gov/oes/currenT/oes435052.htm (accessed: 8 October 2019).

Bureau of Labor Statistics (2018b): Contingent and Alternative Employment Arrangements News Release. Available at: www.bls.gov/news.release/conemp.htm. (accessed: 15 October 2019).

Butts, Marcus M., William J. Becker and Wendy R. Boswell. (2015): Hot buttons and time sinks: The effects of electronic communication during nonwork time on emotions and work-nonwork conflict. *Academy of Management Journal* (58): 763–788.

Cairns, Sally (2005): Delivering supermarket shopping: more or less traffic? *Transport Reviews* 25, No. 1 (January): 51–84.

Calo, R and Rosenblat, A (2017): The taking economy: Uber, information, and power. Columbia Law Review 117(6): 1623–1690.

Campbell, Katherine and Helleloid, Duane (2016): Starbucks: Social responsibility and tax avoidance. *Journal of Accounting Education, Volume* 37, 38–60.

Canaccord Genuity (2015): Apple Claims 92% of Global Smartphone Profits. Available at: www.statista.com/chart/4029/smartphone-profit-share/ (accessed: 6 December 2019).

Canzler, Weert and Andreas Knie (2009): *Grüne Wege aus der Autokrise: Vom Autobauer zum Mobilitätsdienstleister.* Schriften zur Ökologie 4. Berlin: Heinrich-Böll-Stiftung.

Cantarella, Michelle and Strozzi, Chiara (2018): Labour market effects of crowdwork in US and EU: an empirical investigation, Department of Economics, University of Modena and Reggio E.

Canzler, Weert and Andreas Knie (2016): *Die digitale Mobilitätsrevolution: Vom Ende des Verkehrs, wie wir ihn kannten.* Munich: Oekom Verlag.

Cao, Sissi (2019): Apple Refuses to Pay Ireland $14 Billion in Back Taxes—And the Irish Don't Want It. *Observer.* Available at: https://observer.com/2019/09/apple-ireland-tax-lawsuit-european-union-corporate-tax-dodging/ (accessed: 22 November 2019).

Carey, Scott (2017): Alternatives to Uber: The best alternative ride-hailing apps. *techworld. com.* Available at: www.techworld.com/startups/alternatives-uber-best-alternative-ride-hailing-apps-3656813/ (accessed: 3 December 2019).

Carli, James (2016): Oceantop Living in a Seastead – Realistic, Sustainable, and Coming Soon. 10 December. *HuffPost News.* Available at: www.huffingtonpost.com/entry/oceantop-living-in-a-seastead-realistic-sustainable_us_584c595ae4b0016e50430490 (accessed: 6 December 2019).

Castells, Manuel (2001): *The Internet galaxy: reflections on the Internet, business, and society.* Oxford, New York: Oxford University Press.

Castells, Manuel (1996): *The Rise of the Network Society, The Information Age: Economy, Society and Culture Vol. I.* Cambridge, Massachusetts; Oxford, UK: Blackwell.

del Castillo, Michael (2018): Big Blockchain: The 50 Largest Public Companies Exploring Blockchain. *Forbes.* Available at: www.forbes.com/sites/michaeldelcastillo/2018/07/03/big-blockchain-the-50-largest-public-companies-exploring-blockchain/ (accessed: 23 October 2019).

Dunlop, Tim (2017): What is a robot exactly – and how do we make it pay tax? *Guardian* (12 March), sec. Guardian Sustainable Business. Available at: www.theguardian.com/sustainable-business/2017/mar/13/what-is-a-robot-exactly-and-how-do-we-make-it-pay-tax/ (accessed: 23 October 2019).

Chan, Jason, Pikki Fung and Pauline Overeem (2016a): The Poisonous Pearl. Occupational chemical poisoning in the electronics industry in the Pearl River Delta, People's Republic of China. Amsterdam: Good Electronics.

Chan, Jenny, Ngai Pun and Mark Selden (2016b): Dying for an iPhone: the lives of Chinese workers. *chinadialogue.net.* Available at: www.chinadialogue.net/article/show/single/en/8826-Dying-for-an-iPhone-the-lives-of-Chinese-workers (accessed: 3 December 2019).

Chancerel, Perrine, Max Marwede, Nils F. Nissen and Klaus-Dieter Lang (2015): Estimating the quantities of critical metals embedded in ICT and consumer equipment. *Resources, Conservation and Recycling* 98: 9–18.

Charlot, Christophe (2016): *Uberize me.* Bruxelles: Editions Racine.

Charniak, Eugene and Drew McDermott (1985): *Introduction to Artificial Intelligence.* Boston: Addison-Wesley Longman Publishing Co., Inc.

Cheng, Cecilia and Angel Yee-lam Li (2014): Internet addiction prevalence and quality of (real) life: a meta-analysis of 31 nations across seven world regions. *Cyberpsychology, Behavior and Social Networking* 17, Nr. 12 (December): 755–760.

Chen, T. Donna and Kara M. Kockelman (2016): Carsharing's life-cycle impacts on energy use and greenhouse gas emissions. *Transportation Research Part D: Transport and Environment* 47 (August): 276–284.

Chen, L., Mislove, A., Wilson, C. (2016): An empirical analysis of algorithmic pricing on amazon marketplace. *Proceedings of the 25th International Conference on WorldWideWeb,* 1339–1349.

Cheng, C., and Li, A. Y. (2014): Internet addiction prevalence and quality of (real) life: a meta-analysis of 31 nations across seven world regions. *Cyberpsychology, behavior and social networking,* 17(12), 755–760.

China Labor Watch (2016): Apple making big profits but Chinese workers' wage on the slide. Available at: www.chinalaborwatch.org/upfile/2016_08_23/Pegatron-report%20 FlAug.pdf ((accessed: 3 December 2019).

Christl, Wolfie (2017): Corporate Surveillance In Everyday Life. How Companies Collect, Combine, Analyze, Trade, and Use Personal Data on Billions. *Cracked Labs – Institute for Critical Digital Culture* (8 June). Available at: http://crackedlabs.org/en/corporate-surveillance (accessed: 6 December 2019).

Cisco (2016): The Zettabyte Era: Trends and Analysis. White Paper. San Jose, Singapore, Amsterdam.

Cisco (2018): Cisco Visual Networking Index: Forecast and Trends, 2017–2022.

Cohen, Robert and Reginald E. Zelnik (2002): *The Free Speech Movement: Reflections on Berkeley in the 1960s*. Berkeley: University of California Press.

Coroama, Vlad C., Lorenz M. Hilty and Martin Birtel (2012): Effects of Internet-based multiple-site conferences on greenhouse gas emissions. *Telematics and Informatics* 29, No. 4. Green Information Communication Technology (November): 362–374.

Coroama, Vlad and Friedemann Mattern (2019): Digital Rebound – Why Digitalization Will Not Redeem Us Our Environmental Sins. *Proceedings of the 6th international conference on ICT for Sustainability (ICT4S 2019). Lappeenranta, Finland, June 2019.*

Counterpoint Research (2018): Share of mobile phone sales profit by vendor worldwide from 2016 to 2018. *Statista*. Available at: www.statista.com/statistics/780367/global-mobile-handset-profit-share-by-vendor/ (accessed: 19 November 2019).

Cross, Gary S. (2000): *An all-consuming century: Why commercialism won in modern America.* New York: Columbia University Press.

Csikszentmihályi, Christopher (2017): Making A Fresh Start. Event: Die transformative Kraft der Maker, 3 January, Berlin. Available at: www.cowerk.org/veranstaltungen/die-transformative-kraft-der-maker.html (accessed: 6 December 2019).

Dahler, Don (2014): Do companies charge online shoppers different prices?. *CBS*. Available at: www.cbsnews.com/news/do-companies-charge-online-shoppers-different-prices/ (accessed: 11 November 2019).

D'Alisa, Giacomo, Federico Demaria and Giorgos Kallis (2014): *Degrowth: a vocabulary for a new era*. New York, London: Routledge.

van Dam, S. S., Bakker, C. A. and Buiter, J. C. (2013): Do home energy management systems make sense? Assessing their overall lifecycle impact. *Energy Policy* 63 (December): 398–407.

Dämon, Kerstin (2015): Studie Digitalisierung und Arbeitsplätze: Computer können Jobs von 4,4 Millionen Deutschen übernehmen. *Wirtschaftswoche*. 16 December. Available at: www.wiwo.de/erfolg/beruf/studie-digitalisierung-und-arbeitsplaetze-computer-koennen-jobs-von-4-4-millionen-deutschen-uebernehmen/12724850.html (accessed: 28 June 2019).

Daten-Speicherung.de (2017): Minimum Data, Maximum Privacy. Available at: www.daten-speicherung.de/index.php/ueberwachungsgesetze/ (accessed: 8 November 2019).

Daugherty, Paul, Negm, Walid, Banerjee, Prith, Alter, Allan (2015) "Driving Unconventional Growth through the Industrial Internet of Things" (PDF). *Accenture*. Available at: www.accenture.com/us-en/_acnmedia/accenture/next-gen/reassembling-industry/pdf/accenture-driving-unconventional-growth-through-iiot.pdf (accessed: 12 November 2019).

Day, M. and J. Gu (2019): The Enormous Numbers Behind Amazon's Market Reach. *Bloomberg*. Available at: www.bloomberg.com/graphics/2019-amazon-reach-across-markets/ (accessed: 19 November 2019).

de Decker, Kris (2015): Why We Need a Speed Limit for the Internet. *Low-Tech Magazine – Doubts on progress and technology*. 19 October. Available at: www.lowtechmagazine.com/2015/10/can-the-internet-run-on-renewable-energy.html (accessed: 7 July 2019).

Deloitte Touche Tohmatsu Limited (2019): Global Powers of Retailing. *Deloitte*. Available at: www2.deloitte.com/content/dam/Deloitte/global/Documents/Consumer-Business/cons-global-powers-retailing-2019.pdf (accessed: 04 November 2019).

Deutsches Spionagemuseum Berlin (2019): Deutsches Spionagemuseum Berlin – German Spy Museum Berlin. Available at: www.deutsches-spionagemuseum.de/en (accessed: 18 October 2019).

Digiconomics (2019): Bitcoin Energy Consumption. Available at: https://digiconomist.net/bitcoin-energy-consumption (accessed: 23 October 2019).

Dolan, P, T Peasgood and M White (2008): Do we really know what makes us happy? A review of the economic literature on the factors associated with subjective well-being. *Journal of Economic Psychology* 29, No. 1: 94–122.

Domo (2019): Data Never Sleeps 7.0. *Domo*. Available at: www.domo.com/learn/data-never-sleeps-7 (accessed: 10 December 2019).

Druckman, Angela, Ian Buck, Bronwyn Hayward and Tim Jackson (2012): Time, gender and carbon: A study of the carbon implications of British adults' use of time. Ecol. Econ. 84, 153–163.

Duhigg, Charles and David Kocieniewski (2012): Apple's Tax Strategy Aims at Low-Tax States and Nations. *New York Times*. 28 April. Available at: www.nytimes.com/2012/04/29/business/apples-tax-strategy-aims-at-low-tax-states-and-nations.html (accessed: 25 September 2019).

Dusi, Davide (2016): The Perks and Downsides of Being a Digital Prosumer: Optimistic and Pessimistic Approaches to Digital Prosumption. *International Journal of Social Science and Humanity*, Vol. 6, No. 5, 375–381.

Easterlin, Richard A. (1974): Does economic growth improve the human lot? Some empirical evidence. *Nations and households in economic growth* 89: 89–125.

Easterlin, Richard A. and McVey, Laura A (2010): The happiness-income paradox revisited. *Proceedings of the National Academy of Sciences* 107, No. 52: 22463–22468.

EDRi (2019): ePrivacy: EU Member States push crucial reform on privacy norms close to a dead end. European Digital Rights. Available at: https://edri.org/eprivacy-eu-member-states-push-crucial-reform-on-privacy-norms-close-to-a-dead-end/ (accessed: 10 December 2019).

Egger, Nina (2017): Ein neuer Weg: Dezentrale Energieversorgung. *TEC21*, No. 7–8: 26–28.

E-ISAC (2016): Analysis of the Cyber Attack on the Ukrainian Power Grid Defense Use Case. *TLP*: White. Available at: https://ics.sans.org/media/E-ISAC_SANS_Ukraine_DUC_5.pdf (accessed: 4 November 2019).

Elfvengren, Kalle, Matti Karvonen, Kimmo Klemola and Matti Lehtovaara (2014): The future of decentralised energy systems: insights from a Delphi study. *International Journal of Energy Technology and Policy* 10, Nr. 3/4: 265.

Elsberg, Marc (2017): *BLACKOUT: a novel*. Naperville: Sourcebooks Landmark.

EMC (2014): The digital universe of opportunities. Rich data and the increasing value of the internet of things.

eMarketer (2019a): Retail e-commerce sales worldwide from 2014 to 2023. *Statista*. Available at: www.statista.com/statistics/379046/worldwide-retail-e-commerce-sales/ (accessed: 4 November 2019).

eMarketer (2019b): Ausgaben für Online-Werbung weltweit in den Jahren 2013 bis 2018 sowie eine Prognose bis 2023 (in Milliarden US-Dollar). *Statista*. Available at: https://de.statista.com/statistik/daten/studie/185637/umfrage/prognose-der-entwicklung-der-ausgaben-fuer-online-werbung-weltweit/ (accessed: 4 November 2019).

eMarketer Editors (2019): Digital Investments Pay Off for Walmart in Ecommerce Race. *eMarketer*. Available at: www.emarketer.com/content/digital-investments-pay-off-for-walmart-in-ecommerce-race (accessed: 19 November 2019).

EPA (2018): Advancing Sustainable Materials Management: 2015 Fact Sheet. Available at: www.epa.gov/sites/production/files/2018-07/documents/2015_smm_msw_factsheet_07242018_fnl_508_002.pdf (accessed: 11 November 2019).

Ericsson, A. B. (2017): Mobility Report: Traffic Exploration. June. Available at: www.ericsson.com/TET/trafficView/loadBasicEditor.ericsson (accessed: 9 October 2019).

Ethereum (2017): Ethereum. Blockchain App Platform. Available at: https://ethereum.org/ (accessed: 2 November 2017).

European Commission (2012): The Digital Agenda for Europe – Driving European growth digitally. COM (2012) 784 final. Brussels: European Commission.

European Commission (2016): *Regulation of the European Parliament and of the Council on the protection of natural persons with regard to the processing of personal data and on the free movement of such data (General Data Protection Regulation).*

European Commission (2018a): Consumer market study on online market segmentation through personalised pricing/offers in the European Union, European Commission. Available at: https://ec.europa.eu/info/sites/info/files/aid_development_cooperation_fundamental_rights/aid_and_development_by_topic/documents/synthesis_report_online_personalisation_study_final_0.pdf (accessed: 11 November 2019).

European Commission (2018b): Environmental potential of the collaborative economy. Brussels. Available at: https://op.europa.eu/en/publication-detail/-/publication/8e18cbf3-2283-11e8-ac73-01aa75ed71a1/language-en. (accessed: 11 November 2019).

European Digital Rights (EDRi) (2019): Net Neutrality. Available at: https://edri.org/ (accessed: 19 November 2019).

European Network for Rural Development (2018a): Digital Villages Germany: Working document. Available at: https://enrd.ec.europa.eu/sites/enrd/files/tg_smart-villages_case-study_de.pdf (accessed: 25 November 2019).

European Network for Rural Development (2018b): Smart Countryside study Finland: Working document. Available at: https://enrd.ec.europa.eu/sites/enrd/files/tg_smart-villages_case-study_fi_0.pdf (accessed: 25 November 2019).

European Network for Rural Development (2018c): Strategy for Inner Areas Italy: Working document. Available at: https://enrd.ec.europa.eu/sites/enrd/files/tg_smart-villages_case-study_it.pdf (accessed: 26 November 2019).

European Parliament (2010): Decentralized Energy Systems. Directorate General for International Policies. Brussels. Available at: www.europarl.europa.eu/document/activities/cont/201106/20110629ATT22897/20110629ATT22897EN.pdf (accessed: 11 November 2019).

European Parliament (2018): Report on autonomous driving in European transport. Available at: www.europarl.europa.eu/doceo/document/A-8-2018-0425_EN.pdf (accessed: 11 November 2019).

European Parliamentary Research Service (2016): Electricity 'Prosumers'. Briefing. Available at: www.europarl.europa.eu/RegData/etudes/BRIE/2016/593518/EPRS_BRI(2016)593518_EN.pdf (accessed: 1 November 2019).

European Union (2016): *Directive (EU) 2016/681 of the European Parliament and of the Council of 27 April 2016 on the use of passenger name record (PNR) data for the prevention, detection, investigation and prosecution of terrorist offences and serious crime.* Available at: https://eur-lex.europa.eu/legal-content/EN/TXT/?uri=CELEX:32016L0681 (accessed: 6 December 2019).

European Union Agency For Network and Information Security (2017): Privacy and data protection in mobile applications: A study on the app development ecosystem

and the technical implementation of GDPR. Available at: www.enisa.europa.eu/pub
lications/privacy-and-data-protection-in-mobile-applications/at_download/fullReport
(accessed: 23 November 2019).

Eurostat Statistics Explained (2019): Packaging waste statistics. Available at: https://ec.
europa.eu/eurostat/statistics-explained/index.php/Packaging_waste_statistics#Waste_
generation_by_packaging_material (accessed: 11 November 2019).

Evans, Benedict (2017): Cars and second order consequences. Available at: http://ben-
evans.com/benedictevans/2017/3/20/cars-and-second-order-consequences (accessed:
1 August 2019).

Facebook (2019): *Terms of Service*. San Mateo: Facebook. Available at: www.facebook.com/
legal/terms/plain_text_terms (accessed: 8 November 2019).

Fagnant, Daniel J. and Kockelman, Kara (2015): Preparing a nation for autonomous vehicles: opportunities, barriers and policy recommendations. *Transportation Research Part A:
Policy and Practice* 77 (1 July): 167–181.

Fair Crowd Work (2017): Shedding light on the real work of crowd-, platform-, and app-
based work. Available at: http://faircrowd.work/unions-for-crowdworkers/ (accessed: 27
November 2019).

Fawkes, Johanna & Gregory, Anne (2000): Applying communication theories to the Internet. *Journal of Communication Management*, No. 5: 109–124.

Federal Ministry for Economic Affairs and Energy (2019): Securing of Skilled Labour.
Article. Available at: www.bmwi.de/Redaktion/EN/Dossier/skilled-professionals.html
(accessed: 27 November 2019).

Federal Trade Commission (2017): Understanding mobile apps. Available at: www.con
sumer.ftc.gov/articles/0018-understanding-mobile-apps (accessed: 23 November 2019).

Feenstra, Robert C., Inklaar, Robert and Timmer, Marcel P. (2015): The next generation of
the Penn World Table. *The American Economic Review* 105, No. 10: 3150–3182.

Feiner, Lauren (2019): Amazon admits to Congress that it uses 'aggregated' data from
third-party sellers to come up with its own products. CNBC. Available at: www.
cnbc.com/2019/11/19/amazon-uses-aggregated-data-from-sellers-to-build-its-own-
products.html (accessed: 10 December 2019).

Ferdinand, Jan-Peter, Petschow, Ulrich and Dickel, Sascha, Eds (2016): *The Decentralized and
Networked Future of Value Creation*. Cham: Springer International Publishing.

Fichter, K., R. Hintemann, S. Beucker and S. Behrendt (2012): Gutachten zum Thema
"Green IT-Nachhaltigkeit" für die Enquete-Kommission Internet und digitale
Gesellschaft des Deutschen Bundestages.

Foo Yun Chee (2019): Apple spars with EU as $14 billion Irish tax dispute drags on. *Reuters*.
Available at: www.reuters.com/article/us-eu-apple-stateaid/apple-spars-with-eu-as-14-
billion-irish-tax-dispute-drags-on-idUSKBN1W31FE (accessed: 28 October 2019).

Foucault, Michel (1977): *Discipline and Punish: The Birth of the Prison*. New York: Pantheon
Books.

Frey, Carl B. and Osborne, Michael A. (2013): The future of employment: How susceptible are
Jobs to Computerisation? Working Paper. Oxford Martin School, University of Oxford.

Fritzsche, Kerstin, Luke Shuttleworth, Bernhard Brand and Philipp Blechinger (2019):
Exploring the Nexus of Digital Technologies and Mini-grids for Sustainable Energy
Access [IASS Study]. Potsdam: IASS, Reiner Lemoine Institute & Enerpirica.

Galbraith, John Kenneth (1958): *The Affluent Society*. Boston: Houghton Mifflin.

Gates, Bill (1976): An open letter to hobbyists. *Homebrew Computer Club Newsletter* 2, No. 1: 2.

German Advisory Council on Global Change (WBGU), ed. (2011): *World in Transition A Social
Contract for Sustainability*. Berlin: German Advisory Council on Global Change (WBGU).

Gershenfeld, Neil (2012): How to make almost anything: The digital fabrication revolution. *Foreign Affairs.* 91: 43.

Gershgorn, Dave (2016): DRIVING AWAY. The White House predicts nearly all truck, taxi, and delivery driver jobs will be automated. *Quartz.* Available at: https:// qz.com/868716/the-white-house-predicts-nearly-all-truck-taxi-and-delivery-driver-jobs-will-be-automated/ (accessed: 22 October 2019).

GeSI and Accenture (2015): Smarter 2030. ICT Solutions for 21st Century Challenges. Brussels.

GeSI and Deloitte (2019): Digital with Purpose: Delivering a SMARTer2030. Brüssel.

Gibb, Alicia (2014): *Building open source hardware: DIY manufacturing for hackers and makers.* London: Pearson Education.

Gibbs, Samuel (2018): WhatsApp sharing user data with Facebook would be illegal, rules ICO. *Guardian.* Available at: www.theguardian.com/technology/2018/mar/14/ whatsapp-sharing-user-data-facebook-illegal-ico-gdpr (accessed: 25 November 2019).

Ginder *et al.* (2017): Enrollment and Employees in Postsecondary Institutions, Fall 2016; and Financial Statistics and Academic Libraries, Fiscal Year 2016. *RTI International.* Available at: https://nces.ed.gov/pubs2018/2018002.pdf (accessed: 22 October 2019).

Gleick, James (1999): *Faster. The Acceleration of Just About Everything.* New York: Vintage Books.

Goldman Sachs (2016): Virtual and augmented reality: Understanding the race for the next computing platform (13 January). Available at: www.goldmansachs.com/our-thinking/ pages/technology-driving-innovation-folder/virtual-and-augmented-reality/report.pdf (accessed: 6 December 2019).

Goleman, Daniel and Gregory Norris (2010): How Green Is My iPad? *New York Times* (4 April). Available at: https://archive.nytimes.com/www.nytimes.com/interactive/ 2010/04/04/opinion/04opchart.html (accessed: 6 December 2019).

Goodall, N. J. (2016): Can you program ethics into a self-driving car? *IEEE Spectrum,* vol. 53, no. 6, 28–58.

Gorge, Hélène, Maud Herbert, Nil Özcaglar-Toulouse and Isabelle Robert (2015): What Do We Really Need? Questioning Consumption Through Sufficiency. *Journal of Macromarketing* 35, Nr. 1: 11–22.

Gossen, Maike, Sabrina Ludmann and Scholl, Gerd (2017): Sharing is caring – for the environment. Results of life cycle assessments for peer-to-peer sharing. Event: 4th IWSE, Lund.

Gossen, Maike, Florence Ziesemer and Schrader, Ulf (2019): Why and how commercial marketing should promote sufficient consumption: a systematic literature review. *Journal of Macromarketing* 39, Nr. 3 (30 July): 252–269.

Gößling-Reisemann, Stefan (2016): Resilience – Preparing Energy Systems for the Unexpected. *IRGC Resource Guide on Resilience, EPFL International Risk Governance Council.*

Greenfield, Adam (2017): *Radical Technologies: The Design of Everyday Life.* London; New York: Verso.

Greenpeace (2014): *Clicking clean: How companies are creating the green internet.* Washington.

Greenpeace (2017a): Clicking Green. Who Is Winning the Race to Build a Green Internet? Washington.

Greenpeace (2017b): From Smart To Senseless: The Global Impact of 10 Years of Smartphone.

Greenstein, Joshua (2019): The Precariat Class Structure and Income Inequality Among US Workers: 1980–2018. Working Papers. New School for Social Research, Department of Economics.

Greive, Martin, Hildebrand, Jan, Berschens, Ruth (2018): Trump set to tussle over EU digital tax plans. *Handelsblatt*. Available at: www.handelsblatt.com/today/politics/pushing-buttons-trump-set-to-tussle-over-eu-digital-tax-plans/23581852.html (accessed: 28 October 2019).

Greveler, Ulrich & Justus, Benjamin & Loehr, Dennis (2012): Forensic content detection through power consumption. *IEEE International Conference on Communications*.

Groß, Michael (2015): Mobile shopping: a classification framework and literature review. *International Journal of Retail & Distribution Management* 43, No. 3 (18 February): 221–241.

Grothaus, Michael (2018): Amazon is under fire in Germany for destroying as-new and returned items. *Fast Company*. Available at: www.fastcompany.com/40583451/amazon-is-under-fire-in-germany-for-destroying-as-new-and-returned-items (accessed: 11 November 2019).

Gruel, Wolfgang and Stanford, Joseph M. (2016): Assessing the Long-term Effects of Autonomous Vehicles: A Speculative Approach. *Transportation Research Procedia* 13: 18–29.

Guardian (2018): Tech giants may have to be broken up, says Tim Berners-Lee. Available at: www.theguardian.com/technology/2018/nov/01/tim-berners-lee-says-says-tech-giants-may-have-to-be-broken-up (accessed: 25 November 2019).

Gueham, Farid (2017): Digital Sovereignty – Steps towards a new system of internet governance. *The Fondation pour l'innovation politique*. Available at: https://euagenda.eu/publications/digital-sovereignty-steps-towards-a-new-system-of-internet-governance (accessed: 27 November 2019).

Gupta, Richie (2019): Just how far ahead is Alibaba in China's e-Commerce market. *Market Realist*. Available at: https://articles2.marketrealist.com/2019/07/just-how-far-ahead-is-alibaba-in-chinas-e-commerce-market/ (accessed: 22 November 2019).

Gutiérrez, Alfonso and Tyner, Kathleen. (2012): Media Education, Media Literacy and Digital Competence. *Comunicar*. 19.

Habermas, Jürgen (2014): *Ach, Europa*. 5. Aufl. Kleine politische Schriften 11. Frankfurt am Main: Suhrkamp.

Harari, Yuval Noah (2018): *21 lessons for the 21st century*. New York: Spiegel & Grau.

Hazas, Mike, Morley, Janine, Bates Oliver and Friday, Adrian (2016): Are there limits to growth in data traffic?: on time use, data generation and speed. *Proceedings of the Second Workshop on Computing Within Limits*, pp. 14:1–14:5. LIMITS '16. New York, NY, USA.

Heinberg, R (2011): *The end of growth: Adapting to our new economic reality*. Gabriola Island: New Society Publishers.

Helbing, Dirk, Frey, Bruno S., Gigerenzer, Gerd, Hafen, Ernst, Hagner, Michael, Hofstetter, Yvonne, Zicari Roberto V. and Zwitter, Andrej (2017): Will Democracy Survive Big Data and Artificial Intelligence? *Scientific American* (25 February). Available at: www.scientificamerican.com/article/will-democracy-survive-big-data-and-artificial-intelligence/. (accessed: 30 November 2019).

Hermann, Christoph, Christopher Schmidt, Dennis Kurle, Stefan Blume and Sebastian Thiede (2014): Sustainability in manufacturing and factories of the future. *International Journal of Precision Engineering and Manufacturing-Green Technology* 1, No. 4: 283–292.

Hermant, Bertrand (2019): INSIGHT: France Taxes the Digital Economy. *Bloomberg Tax*. Available at: https://news.bloombergtax.com/daily-tax-report-international/insight-france-taxes-the-digital-economy (accessed: 28 October 2019).

Hern, A. and Waterson, J, (2019): Social media firms fight to delete Christchurch shooting footage. *Guardian*. Available at: www.theguardian.com/world/2019/mar/15/video-of-christchurch-attack-runs-on-social-media-and-news-sites (accessed: 22 November 2019).

Hielscher, H., Goebel, J., and Brück, M. (2018): Amazon destroys massive quantities of returned and as-new goods. *WirtschaftsWoche Online*. Available at: www.wiwo.de/unternehmen/handel/online-retailer-amazon-destroys-massive-quantities-of-returned-and-as-new-goods/22662746.html (accessed: 11 November 2019).

Hillebrand, A., Thiele, S., Junk, P., Hildebrandt, C., Needham, P., Kortüm, M. (2016): Technology and change in postal services – impacts on consumers. Bad Honnef: *Wik Consult*. p. 54–58. Available at: www.wik.org/fileadmin/Studien/2016/WIK-Consult_CitA_Impact_of_technology_full-report.pdf (accessed: 8 October 2019).

Hilty, Lorenz M. (2008): *Information technology and sustainability: essays on the relationship between ICT and sustainable development*. Norderstedt: Books on Demand.

Hilty, Lorenz M. (2012): Why energy efficiency is not sufficient – some remarks on «Green by IT». *EnviroInfo 2012*, ed. by H.K. Arndt, 13–20.

Hilty, Lorenz M. and Aebischer, Bernard, Eds (2015a): *ICT Innovations for Sustainability*. Vol. 310. Advances in Intelligent Systems and Computing. Cham: Springer International Publishing.

Hilty, Lorenz M. and Aebischer, Bernard (2015b): ICT for Sustainability: An Emerging Research Field. *ICT Innovations for Sustainability*, ed. by Lorenz M. Hilty and Bernard Aebischer, 310: pp. 3–36. Cham: Springer International Publishing.

Hilty, Lorenz M., Wolfgang Lohmann, Siegfried Behrendt, Michaela Evers-Wölk, Klaus Fichter and Ralph Hintemann (2015): Green Software. Final report of the project: Establishing and exploiting potentials for environmental protection in information and communication technology (Green IT). Dessau-Roßlau: Federal Environment Agency (Germany).

Hines, Colin (2000): *Localization: A Global Manifesto*. London: Routledge.

Hines, Nickolaus (2016): This Internet Map Shows our "Connected Worldview." 8 August. *Inverse*. Available at: www.inverse.com/article/19357-internet-map-of-every-connected-device (accessed: 3 November 2019).

Hintemann, Ralph and Clausen, Jens (2016): Green Cloud? The current and future development of energy consumption by data centers, networks and end-user devices, pp. 109–115.

Hirsch, Fred (1995): *Social limits to growth*. revised. London: Routledge.

Hobsbawm, Eric. J. (1952): The machine breakers. *Past & Present*, No. 1: 57–70.

Hochschorner, Elisabeth, Dán, György and Moberg, Åsa (2015): Carbon footprint of movie distribution via the internet: a Swedish case study. *Journal of Cleaner Production* 87 (15 January): 197–207.

Höfner, Anja and Santarius, Tilman (2017): Wertschätzungs- statt Wegwerfgesellschaft. Soziale Innovation dank Digitalisierung. *Politische Ökologie* 150: 139–144.

Hogan, K. (2018): Consumer Experience in the Retail Renaissance: How Leading Brands Build a Bedrock with Data. *Deloitte*. Available at: www.deloittedigital.com/us/en/blog-list/2018/consumer-experience-in-the-retail-renaissance-how-leading-brand.html (accessed: 11 November 2019).

Hogg, Nick and Jackson, Tim (2009): Digital Media and Dematerialization: An Exploration of the Potential for Reduced Material Intensity in Music Delivery. *Journal of Industrial Ecology* 13, No. 1 (February): 127–146.

Holzer, J.M., M.C. Adamescu, F.J. Bonet-García, R.. Díaz Delgado, J. Dick, J.M. Grove, R. Rozzi and D.E. Orenstein (2018): Negotiating local versus global needs in the International Long Term Ecological Research Network's socio-ecological research agenda. *Environmental Research Letters* No. 13 (2018).

Horner, Nathaniel C., Arman Shehabi and Inês L. Azevedo. (2016): Known unknowns: indirect energy effects of information and communication technology. *Environmental Research Letters* 11, No. 10: 103001.

Horowitz Research (2016): *State of Cable & Digital Media*. New York: Horowitz Research.

Howcroft, D. and Bergvall-Kåreborn, B. (2019): A Typology of Crowdwork Platforms. *Work, Employment and Society*, 33(1), 21–38.

Hudson, Michael (2017): The Road to Debt Deflation, Debt Peonage, and Neofeudalism. Levy Economics Institute of Bard College Working Paper No. 708. Available at: http://dx.doi.org/10.2139/ssrn.2007284.

Huhn, Philipp (2016): Smart Home – Energy Management Outlook. Hamburg: statista GmbH.

Huws, Ursula, Neil H. Spencer and Simon Joyce (2016): Crowd Work in Europe: Preliminary results from a survey in the UK, Sweden, Germany, Austria and the Netherlands. FEPS Studies.

ICTechX (2019): RFID Forecasts, Players and Opportunities 2019–2029. Available at: www.idtechex.com/en/research-report/rfid-forecasts-players-and-opportunities-2019-2029/700 (accessed: 13 November 2019).

Illich, Ivan (1973): *Tools for conviviality*. New York: Harper & Row.

Interactive Advertising Bureau (2014): Mobile Location Use Cases and Case Studies. A look at current implementation and best practices for mobile location data. Available at: www.iab.com/wp-content/uploads/2015/07/MobileLocationUseCasesandCaseStudies Final.pdf (accessed: 5 November 2019).

International Energy Agency, Ed. (2016): *World energy outlook 2016*. Paris: OECD.

International Federation of Robotics (2017): The Impact of Robots on Productivity, Employment and Jobs. Available at: https://ifr.org/img/office/IFR_The_Impact_of_Robots_on_Employment.pdf (accessed: 18 October 2019).

International Labour Organization, Ed. (2013): *Wages and equitable growth*. Global wage report [3.]2012/13. Geneva.

International Labour Organization (2015): The LabourShare in G20Economies. Report prepared for the G20 Employment Working Group Antalya, Turkey. Available at: www.oecd.org/g20/topics/employment-and-social-policy/The-Labour-Share-in-G20-Economies.pdf (accessed: 18 October 2019).

International Labour Organization (2019): ILOSTAT The leading source of labour statistics. Available at: www.ilo.org/shinyapps/bulkexplorer17/?lang=en&segment=indicator&id=SDG_1041_NOC_RT_A (accessed: 6 December 2019).

Ipakchi, A. and Albuyeh, F. (2009): Grid of the future. *IEEE Power and Energy Magazine*, vol. 7, no. 2, 52–62.

IPCC (2014): *Climate change 2014: synthesis report*. Ed. by Core Writing Team, Rajendra Kumar Pachauri, and Leo Meyer. Geneva, Switzerland.

IPCC (2018): *Global Warming of 1.5°C. An IPCC Special Report on the impacts of global warming of 1.5°C above pre-industrial levels and related global greenhouse gas emission pathways, in the context of strengthening the global response to the threat of climate change, sustainable development, and efforts to eradicate poverty*. Available at: www.ipcc.ch/site/assets/uploads/sites/2/2019/06/SR15_Full_Report_Low_Res.pdf (accessed: 1 November 2019).

Ipeirotis, P., Little, G. and Malone, T.W. (2014): Composing and analyzing crowdsourcing workflows. Collective Intelligence. *Panos Ipeirotis* Available at: www.ipeirotis.com/wp-content/uploads/2014/03/main.pdf (accessed: 15 October 2019).

Jackson, Tim (2011): *Prosperity without growth: economics for a finite planet*. London: Earthscan.

Jackson, Tim (2016): *Prosperity without Growth: Economics for a Finite Planet*. 2nd Revised Edition. London: Taylor & Francis Ltd.

Jarrahi, Mohammad Hossein and Will Sutherland (2019): Algorithmic Management and Algorithmic Competencies: Understanding and Appropriating Algorithms in Gig Work. *Information in Contemporary Society*, ed. Natalie Greene Taylor, Caitlin Christian-Lamb, Michelle H. Martin, and Bonnie Nardi, 11420: 578–589. Cham: Springer International Publishing.

Jeswani, Harish K. and Azapagic, Adisa (2015): Is e-reading environmentally more sustainable than conventional reading? *Clean Technologies and Environmental Policy* 17, No. 3 (March): 803–809.

Jewell, Catherine (2016): Digital pioneer, Jaron Lanier, on the dangers of "free" online culture. *World Intellectual Property Organization*. Available at: www.wipo.int/wipo_magazine/en/2016/02/article_0001.html (accessed: 11 July 2019).

Joumaa, Chibli and Seifedine Kadry (2012): Green IT: Case Studies. *Energy Procedia* 16: 1052–1058.

Jung, Jiyeon and Yoonmo Koo (2018): Analyzing the Effects of Car Sharing Services on the Reduction of Greenhouse Gas (GHG) Emissions. *Sustainability* 10, Nr. 2 (17 February): 539.

Kaldor, Nicholas (1957): A model of economic growth. *The Economic Journal* 67, No. 268: 591–624.

Karabarbounis, L. and Neiman, B. (2014): The Global Decline of the Labor Share. *The Quarterly Journal of Economics* 129, No. 1 (1 February): 61–103.

Kaye, Kate (2017): 3 Mini Case Studies Show How Location Data Is Moving Marketing. *Ad Age*. Available at: https://adage.com/article/print-edition/case-s/308190 (accessed: 5 November 2019).

Keeley, Brian (2015): *Income Inequality: The Gap between Rich and Poor*. Paris: OECD Publishing.

Kelly, Kevin (2016): The untold story of magic leap, the world's most secretive startup. *wired.com* (26 April). Available at: www.wired.com/2016/04/magic-leap-vr/(accessed 6 December 2019).

Kelly, John and François, Camille (2018): This is what filter bubbles actually look like. *MIT Technology Review*. Available at: www.technologyreview.com/s/611807/this-is-what-filter-bubbles-actually-look-like/ (accessed: 23 October 2019).

Kern, Eva, Lorenz M. Hilty, Achim Guldner, Yuliyan V. Maksimov, Andreas Filler, Jens Gröger and Stefan Naumann (2018): Sustainable software products—Towards assessment criteria for resource and energy efficiency. *Future Generation Computer Systems* 86 (September): 199–210.

Keutel, Klara (2015): Elektronikartikel: Wochentage mit den besten Preisen. *Faz.net*. Available at: www.faz.net/aktuell/finanzen/meine-finanzen/geld-ausgeben/elektronikartikel-wochentage-mit-den-besten-preisen-13430224.html (accessed: 29 July 2019).

Keppner, Benno, Kahlenborn, Walter, Richter, Stephan, Jetzke, Tobias; Lessmann, Antje, Bovenschulte, Marc (2018): Focus on the future: 3D printing Trend report for assessing the environmental impacts. *Umweltbundesamt*. Available at: www.umweltbundesamt.de/sites/default/files/medien/1410/publikationen/fachbroschuere_3d_en_2018-07-04.pdf (accessed: 16 October 2019).

Keynes, John Maynard (1936/2003): The General Theory of Employment, Interest, and Money. Macmillan Cambridge University Press, Preface. Available at: https://cas2.umkc.edu/economics/people/facultypages/kregel/courses/econ645/winter2011/general theory.pdf (accessed: 26 November 2019).

Kharpal, Arjun (2017): Bill Gates wants to tax robots, but the EU says, 'no way, no way'. *CNBC*, Tech Transformer. Available at: www.cnbc.com/2017/06/02/bill-gates-robot-tax-eu.html (accessed: 26 November 2019).

Kittner, Noah, Felix Lill and Daniel M. Kammen (2017): Energy storage deployment and innovation for the clean energy transition. *Nature Energy* 2, No. 9 (31 July): 17125.

Kleber, Claus and Angela Andersen (2016): ZDF-Reportage: Silicon Valley – Zu Besuch bei den Herren der Welt. *ZDF*. Mainz. Available at: www.youtube.com/watch?v=Bmx k0D2EsJI (accessed 6 December 2019).

Klemm, Thomas (2016): Stressiger Beruf: Paketbote für einen Tag. *Faz.net*. 6 December. Available at: www.faz.net/aktuell/beruf-chance/arbeitswelt/stressiger-beruf-paketbote-fuer-einen-tag-14557795.html (accessed: 28 June 2019).

Kohtala, Cindy (2017): Making "Making" Critical: How Sustainability is Constituted in Fab Lab Ideology, *The Design Journal*, 20:3, 375–394.

Koj, J. C., Wulf, C. and Zapp, P. (2019): Environmental impacts of power-to-X systems – A review of technological and methodological choices in Life Cycle Assessments. *Renewable and Sustainable Energy Reviews*, Volume 112, 865–879.

Koomey, Jonathan, Berard, Stephen, Sanchez, Marla and Wong, Henry (2011): Implications of historical trends in the electrical efficiency of computing. *IEEE Annals of the History of Computing* 33, No. 3: 46–54.

Kopp, Thomas, Becker, Maximilian, Decker, Samuel, Eicker, Jannis, Engelmann, Hannah, Eradze, Ia, Forster, Franziskus, Haller, Stella, Heuwieser, Magdalena, Hoffmann, Maja et al. (2019): *At the Expense of Others? How the imperial mode of living prevents a good life for all.* Munich: oekom.

Kozak, Greg L. and Keolelan, G. A. (2003): Printed scholarly books and e-book reading devices: a comparative life cycle assessment of two book options. *IEEE International Symposium on Electronics and the Environment, 2003*, 291–296.

Kreiger, M.A., M.L. Mulder, A.G. Glover and J.M. Pearce (2014): Life cycle analysis of distributed recycling of post-consumer high density polyethylene for 3-D printing filament. *Journal of Cleaner Production* 70 (May): 90–96.

Krzanich, Brian (2015): Data is the New Oil in the Future of Automated Driving. *Intel Newsroom*. 15 November. Available at: https://newsroom.intel.com/editorials/krzanich-the-future-of-automated-driving/ (accessed: 7 August 2019).

Kuzyakov, Evgeny and Pio, David (2016): Next-generation video encoding techniques for 360 video and VR. 21 January. *Facebook*. Available at: https://code.facebook.com/posts/1126354007399553/next-generation-video-encoding-techniques-for-360-video-and-vr/(accessed: 6 December 2019).

Kuss, Daria and Griffiths, Mark (2017): Social Networking Sites and Addiction: Ten Lessons Learned. *International Journal of Environmental Research and Public Health*. DOI:14.10.3390/ijerph14030311.

Lange, Steffen (2018): *Macroeconomics Without Growth: Sustainable Economies in Neoclassical, Keynesian and Marxian Theories.* Marburg: Metropolis.

Lange, Steffen (2019): Beyond A-Growth: Sustainable Zero Growth. In *Handbook of Global Sustainability Governance.* pp. 222–233. London: Routledge.

Lange, Steffen, Maximilian Banning, Anne Berner, Florian Kern, Christian Lutz, Jan Peuckert, Tilman Santarius and Alexander Silbersdorff (2019): *Economy-Wide Rebound Effects: State of the art, a new taxonomy, policy and research gaps.* Discussion Paper. Berlin: Institut für ökologische Wirtschaftsforschung.

Lange, Steffen, Peter Pütz and Thomas Kopp (2018): Do Mature Economies Grow Exponentially? *Ecological Economics* 147 (May): 123–133.

Lange, Steffen and Jackson, Tim (2019): Speed up the research and realization of growth independence. Ökologisches Wirtschaften, 34(1).

Lanier, Jaron (2018): Ten arguments for deleting your social media accounts right now. Bodley Head.

Latouche, Serge (2009): *Farewell to growth*. Transl. by David Macey. Cambridge: Polity Press.

Lederman, Doug (2018): Who Is Studying Online (and Where). *Inside Higher Ed*. Available at: www.insidehighered.com/digital-learning/article/2018/01/05/new-us-data-show-continued-growth-college-students-studying (accessed: 8 October 2019).

Lee, M., K., Kusbit, D., Metsky, E. and Dabbish, L. (2015): Working with machines: the impact of algorithmic and data-driven management on human workers. *Proceedings of the 33rd annual ACM conference on human factors in computing systems*, Seoul, Republic of Korea.

Leimeister, Jan Marco, Durward, David and Zogaj, Shkodran (2016): Crowd Worker in Deutschland: Eine empirische Studie zum Arbeitsumfeld auf externen Crowdsourcing-Plattformen. Düsseldorf: Hans-Böckler-Stiftung.

Leslie, Stuart W. (1993): *The Cold War and American science: The military-industrial-academic complex at MIT and Stanford*. Columbia University Press.

Levy, Steven (2010): *Hackers: Heroes of the Computer Revolution*. Sebastopol: O'Reilly Media, Inc.

Levy, Steven (2012): Can an Algorithm Write a Better News Story Than a Human Reporter? *Wired*. 24 April. Available at: www.wired.com/2012/04/can-an-algorithm-write-a-better-news-story-than-a-human-reporter/ (accessed: 28 June 2019).

Lewis, Dyani (2016): Will the internet of things sacrifice or save the environment? *Guardian*. 11 December. Available at: www.theguardian.com/sustainable-business/2016/dec/12/will-the-internet-of-things-sacrifice-or-save-the-environment (accessed: 31 October 2019).

Linder, S. B. (1970): The harried leisure class. Columbia University Press, New York.

Lin, Ying (2019): 10 eBay Statistics You Need to Know in 2019 [Infographic]. *Oberlo*. 8 October. Available at: www.oberlo.com/blog/ebay-statistics (accessed: 5 December 2019).

Liu, Zhi-Chao & Jiang, Qiuhong & Zhang, Yang & Li, Tao & Zhang, Hong-Chao (2016): Sustainability of 3D Printing: A Critical Review and Recommendations. 10.1115/MSEC2016-8618.

Lobb, Alison, and Stack, Robert (2019): G20/OECD work: Maybe way beyond digital. *International Tax Review*. Available at: www.internationaltaxreview.com/article/b1h99x 7k2cq5gz/g20oecd-work-maybe-way-beyond-digital (accessed: 28 October 2019).

van Loon, Patricia, Deketele, Lieven, Dewaele, Joost, McKinnon, Alan and Rutherfor, Christine (2015): A comparative analysis of carbon emissions from online retailing of fast moving consumer goods. *Journal of Cleaner Production* 106 (November): 478–486.

Lopez, Sarah (2019): The 3 Best Food Sharing Apps. *FamilyApp*. Available at: https://family app.com/try-a-food-sharing-app/ (accessed: 27 November 2019).

Louis, Jean-Nicolas, Calo, Antonio, Leiviskä, Kauko and Pongrácz, Eva (2015): Environmental Impacts and Benefits of Smart Home Automation: Life Cycle Assessment of Home Energy Management System. *IFAC-PapersOnLine* 48, No. 1: 880–885.

Lucas, Henry C. (2012): *The search for survival*. Santa Barbara, Calif.: Praeger.

Ludmann, Sabrina (2018): Ökologie des Teilens. Bilanzierung der Umweltwirkungen des Peer-to-Peer Sharing. PeerSharing Arbeitsbericht 8. Ifeu, Heidelberg. Available at: www. ifeu.de/wp-content/uploads/Oekologie_des_Teilens_Arbeitspapier_8_.pdf (accessed: 28 November 2019).

Lutter, Stephan, Giljum, Stefan, Lieber, Mirko and Manstein, Christopher (2016): Die Nutzung natürlicher Ressourcen. Bericht für Deutschland 2016. Dessau-Roßlau: Umweltbundesamt.

Malmodin, Jens and Coroama, Vlad (2016): Assessing ICT's enabling effect through case study extrapolation — The example of smart metering. Event: 2016 Electronics Goes Green 2016+ (EGG), September, Berlin.

Malmodin, Jens, Moberg, Åsa, Lundén, Dag, Finnveden, Göran and Lövehagen, Nina (2010): Greenhouse Gas Emissions and Operational Electricity Use in the ICT and Entertainment & Media Sectors. *Journal of Industrial Ecology* 14, No. 5: 770–790.

Mangiaracina, Riccardo, Marchet, Gino, Perotti, Sara and Tumino, Angela (2015): A review of the environmental implications of B2C e-commerce: a logistics perspective. *International Journal of Physical Distribution & Logistics Management* 45, No. 6 (16 June): 565–591.

Manhart, Andreas, Blepp, Markus, Fischer, Corinna, Graulich, Kathrin, Prakash, Siddharth, Priess, Rasmus, Schleicher, Tobias and Tür, Maria (2016): Resource Efficiency in the ICT Sector. Final Report. Hamburg: Greenpeace.

Mansor Iqbal (2019): WeChat Revenue and Usage Statistics. *BusinessofApps*. Available at: www.businessofapps.com/data/wechat-statistics/#1 (accessed: 19 November 2019).

Manzini, Ezio (2013): Resilient systems and cosmopolitan localism – The emerging scenario of the small, local, open and connected space. Schneidewind, U., Santarius, T., Humburg, A. (ed.): *Economy of Sufficiency: Essays on wealth in diversity, enjoyable limits and creating commons*. Chapter 3, p. 77.

Marscheider-Weidemann, Frank, Langkau, Sabine, Hummen, Torsten, Erdmann, Lorenz Luis, Tercero Espinoza, Alberto, Angerer, Gerhard, Marwede, Max and Benecke, Stephan (2016): *Rohstoffe für Zukunftstechnologien 2016: Auftragsstudie*. DERA Rohstoffinformationen 28. Berlin: Fraunhofer Institut für System- und Innovationsforschung.

Marshall, Aarian (2017): Google Maps Adds Location Sharing, Quietly Drools Over Your Data. *Wired*. Available at: www.wired.com/2017/03/google-maps-share-location/ (accessed: 24 November 2019).

Marx, Karl (1969): Die deutsche Ideologie. *MEW 3–1845–1846*. MEW 3. Berlin: Dietz.

Mason, Paul (2015): *PostCapitalism: A Guide to our Future*. New York: Farrar, Strauss and Giroux.

Mattioli, Dana (2012): On Orbitz, Mac Users Steered to Pricier Hotels. *The Wall Street Journal*. Available at: www.wsj.com/articles/SB10001424052702304458604577488822667325882 (accessed: 11 November 2019).

Matsuda, Michiko and Fumihiko Kimura (2015): Usage of a digital eco-factory for sustainable manufacturing. *Cirp Journal of Manufacturing Science and Technology* 9 (May): 97–106.

Mazzucato, Mariana (2018): Let's make private data into a public good. *MIT Technology Review*. Available at: www.technologyreview.com/s/611489/lets-make-private-data-into-a-public-good/ (accessed: 25 November 2019).

McKibben, Bill (2007): *Deep economy: the wealth of communities and the durable future*. New York: Times Books.

McKinsey (2015): The Internet of Things: Mapping the Value Beyond the Hype. McKinsey Global Institute.

McKinsey Global Institute (2019): Digital India: Technology to transform a connected nation. Report. *Mckinsey*. Available at: www.mckinsey.com/business-functions/mckinsey-digital/our-insights/digital-india-technology-to-transform-a-connected-nation (accessed: 27 November 2019).

McQuivey, James (2013): *Digital disruption: unleashing the next wave of innovation*. Las Vegas: Amazon publ.

Meadows, Donella H., Meadows, Dennis L., Zahn, Erich K.O. and Milling, Peter (1972): *Limits to growth: A Report for the Club of Rome's Project on the Predicament of Mankind.* New York: New American Library.

Meadows, Donella H., Randers, Jørgen and Meadows, Dennis L. (2004): *The limits to growth: the 30-year update.* White River Junction: Chelsea Green Publishing Company.

Merchant, Brian (2015): Fully automated luxury communism. *Guardian.* 18 March. Available at: www.theguardian.com/sustainable-business/2015/mar/18/fully-automated-luxury-communism-robots-employment (accessed: 25 September 2019).

Messenger, Jon (2018): Working time and the future of work. *ILO Future of work research paper series.* Available at: www.ilo.org/wcmsp5/groups/public/-dgreports/-cabinet/documents/publication/wcms_649907.pdf (accessed: 27 November 2019).

Milakis, Dimitris, van Arem, Bart and van Wee, Bert (2017): Policy and society related implications of automated driving: A review of literature and directions for future research. *Journal of Intelligent Transportation Systems* 21, No. 4 (4 July): 324–348.

Milakis, Dimitris, Snelder, Maaike, Van Arem, Bart, Van Wee, Bert and Homem de Almeida Correia, Goncalo (2015): Development of automated vehicles in the Netherlands: scenarios for 2030 and 2050. *Delft University of Technology.*

Mills, Mark P. (2013): The Cloud Begins With Coal. Big Data, Big Networks, Big Infrastructures, and Big Power. An Overview of the Electricity Used by the Global Digital Ecosystem. National Mining Association. American Coalition for Clean Coal Electricity.

Mitchell, Stacey (2017): Amazon Is Trying to Control the Underlying Infrastructure of Our Economy. *Vice* Available at: https://motherboard.vice.com/en_us/article/7xpgvx/amazons-is-trying-to-control-the-underlying-infrastructure-of-our-economy (accessed: 25 November 2019).

Moberg, Åsa, Johansson, Martin, Finnveden, Göran and Jonsson, Alex (2010): Printed and tablet e-paper newspaper from an environmental perspective — A screening life cycle assessment. *Environmental Impact Assessment Review* 30, No. 3 (April): 177–191.

Mooney, P., ETC Group (2018): Blocking the chain: Industrial food chain concentration, Big Data platforms and food sovereignty solutions. *ETC Group.* Available at: www.etcgroup.org/sites/www.etcgroup.org/files/files/blockingthechain_english_web.pdf (accessed: 25 November 2019).

Morgan, Blake (2018): Will There Be A Physical Retail Store In 10–20 Years? *Forbes* (15 October). Available at www.forbes.com/sites/blakemorgan/2018/10/15/will-there-be-a-physical-retail-store-in-10-20-years/#1164ec91723f. (accessed: 25 November 2019).

Morozov, Evgeny (2011): *The net delusion: the dark side of internet freedom.* 1st ed. New York: Public Affairs.

Morozov, Evgeny (2013): *To save everything, click here: the folly of technological solutionism.* New York: Public Affairs.

Morozov, Evgeny (2016): Tech titans are busy privatising our data. *Guardian.* Available at: www.theguardian.com/commentisfree/2016/apr/24/the-new-feudalism-silicon-valley-overlords-advertising-necessary-evil (accessed: 22 November 2019).

Muntaner, Carles (2018): Digital Platforms, Gig Economy, Precarious Employment, and the Invisible Hand of Social Class. *International Journal of Health Services*, 48(4), 597–600.

Muro, Mark, Maxim, Robert and Whiton, Jacob (2019): Automation and Artificial Intelligence: How machines are affecting people and places. *Brookings.* Available at: www.brookings.edu/research/automation-and-artificial-intelligence-how-machines-affect-people-and-places/ (accessed: on 22 October 2019).

Naumann, Stefan, Dick, Markus, Kern, Eva and Johann, Timo (2011): The GREENSOFT Model: A reference model for green and sustainable software and its engineering. *Sustainable Computing: Informatics and Systems* 1, Nr. 4 (1 December): 294–304.

New Economics Foundation (2009): The Great Transition. A tale of how it turned out right. London.

New York Times (2012): 'Double Irish With a Dutch Sandwich'. *New York Times*. 28 April. Available at: www.nytimes.com/interactive/2012/04/28/business/Double-Irish-With-A-Dutch-Sandwich.html?_r=0 (accessed: 25 September 2019).

Nova-Institute for Ecology and Innovation (2018): High-tech strategies for small farmers and organic farming. Press Release. *Bio-based News*. Available at: http://news.bio-based.eu/media/2018/06/18-07-02-PR-High-Tech-Organic-Farming.pdf (accessed: 19 November 2019).

Norton, Andrew (2017a): Automation, the changing world of work, and sustainable development. *The International Institute for Environment and Development (IIED)*. Available at: www.iied.org/automation-changing-world-work-sustainable-development (accessed: 22 October 2019).

Norton, Andrew (2017b): Automation and inequality: the changing world of work in the global South. Issue Paper: *The International Institute for Environment and Development (IIED)*. Available at: www.iied.org/automation-changing-world-work-sustainable-development (accessed: 22 October 2019).

Nübold, Wolfgang (2017): "Virtuelle Batterie" der TRIMET wird Teil der KlimaExpo. NRW. *Trimet*. 13 January. Available at: www.trimet.eu/de/presse/pressemitteilungen/2017/2017-01-13-virtuelle-batterie-der-trimet-wird-teil-der-klimaexpo.nrw.

Nyborg, Sophie and Røpke, Inge (2011): Energy impacts of the smart home-conflicting visions. *Energy Efficiency First: The Foundation of a low-carbon Society*, 1849–1860.

OECD (2015): *In It Together: Why Less Inequality Benefits All*. Paris: Organisation for Economic Co-operation and Development.

OECD (2017): Aid for Trade at a Glance 2017: Promoting Trade, Inclusiveness and Connectivity for Sustainable Development. Paris. Available at: https://dx.doi.org/10.1787/aid_glance-2017-graph70-en.

OECD (2018): Personalised Pricing in the Digital Era. Available at: https://one.oecd.org/document/DAF/COMP(2018)13/en/pdf (accessed: 5 November 2019).

Oehme, Ines, Jacob, Anett, Cerny, Lisa, Golde, Michael, Krause, Susann, Löwe, Unnerstall, Christian, Unnerstall, Herwig and Fabian, Matthias (2017): *Strategies against obsolescence: Ensuring a minimum product lifetime and improving product service life as well as consumer information*. Dessau-Roßlau: German Environment Agency.

Office for National Statistics (2019): Which occupations are at highest risk of being automated? 25 March. Available at: www.ons.gov.uk/employmentandlabourmarket/peopleinwork/employmentandemployeetypes/articles/whichoccupationsareathighestriskofbeingautomated/2019-03-25 (accessed: 4 November 2019).

Open Source Ecology (2017): Open Source Ecology. Available at: http://opensourceecology.org (accessed: 8 November 2019).

OpenSourceSeeds (2017): OpenSourceSeeds. Available at: www.opensourceseeds.org/en/ (accessed: 29 November 2019).

O'Neill, Daniel W., Fanning, Andrew L., Lamb, William F. and Steinberger, Julia K. (2018): A good life for all within planetary boundaries. *Nature Sustainability* 1, Nr. 2 (February): 88–95.

O'Regan, Gerard (2012): *A Brief History of Computing*. London: Springer London.

Orlowski, Andrew (2016): Jaron Lanier: Big Tech is worse than Big Oil. 22 April. Available at: www.theregister.co.uk/2016/04/22/jaron_lanier_on_ip/ (accessed: 11 July 2019).

Ostrom, Elinor (1990): *Governing the Commons: The Evolution of Institutions for Collective Action*. Cambridge: Cambridge University Press.

Ottmann, Henning (2006): Liberal, Republican and Deliberative Democracy. Synthesis Philosophica 21 (2): 315–325.

Oxford Economics (2019): How robots change the world. What automation really means for jobs and productivity. Available at: http://resources.oxfordeconomics.com/how-robots-change-the-world (accessed: 22 October 2019).

Palfrey, John and Gasser, Urs (2008): *Born Digital: Understanding the First Generation of Digital Natives*. New York: Basic Books.

Paris Innovation Review (2016): Agriculture and food: the rise of digital platforms. Available at: http://parisinnovationreview.com/articles-en/agriculture-and-food-the-rise-of-digital-platforms (accessed: 19 November 2019).

Pariser, Eli (2011): *The filter bubble: what the Internet is hiding from you*. New York, NY: Penguin Press.

Patrizio, Andy (2017): 35 Blockchain Startups to Watch. *Datamation*. Available at: Available at: www.datamation.com/data-center/35-blockchain-startups-to-watch.html (accessed: 2 November 2019).

Pawel, Miriam (2019): You Call It the Gig Economy. California Calls It 'Feudalism.' *New York Times*. Available at: www.nytimes.com/2019/09/12/opinion/california-gig-economy-bill-ab5.html (accessed: 22 November 2019).

Pesce, M., Kirova, M., Soma, K., Bogaardt, M.-J., Poppe, K., Thurston, C., Monfort Belles, C, Wolfert, S., Beers, G. and Urdu, D. (2019): Research for AGRI Committee – Impacts of the digital economy on the food-chain and the CAP. European Parliament, Policy Department for Structural and Cohesion Policies, Brussels. Available at: www.europarl.europa.eu/RegData/etudes/STUD/2019/629192/IPOL_STU(2019)629192_EN.pdf (accessed: 25 November 2019).

Petschow, Ulrich, Ferdinand, Jan-Peter, Dickel, Sascha, Flämig, Heike and Steinfeldt, Michael, Eds (2014): *Dezentrale Produktion, 3D-Druck und Nachhaltigkeit: Trajektorien und Potenziale innovativer Wertschöpfungsmuster zwischen Maker-Bewegung und Industrie 4.0*. Schriftenreihe des IÖW 206. Berlin: Institut für ökologische Wirtschaftsforschung.

Petschow, U., Lange, S., Hoffmann, D., Pissarskoi, E., aus dem Moore, N., Korfhage, T. and Schoofs, A. (2018): Social well-being within planetary boundaries. Environmental Research of the Federal Ministry for the Environment, Nature Conservation and Nuclear Safety.

Pew Research Center (2015): Apps Permissions in the Google Play Store. Available at: www.pewresearch.org/internet/2015/11/10/apps-permissions-in-the-google-play-store/ (accessed: 23 November 2019).

Pfeiffer, S. (2017): The Vision of "Industrie 4.0" in the Making—a Case of Future Told, Tamed, and Traded. *NanoEthics*.

Piketty, Thomas (2014): *Capital in the Twenty-First Century*. Cambridge: Harvard University Press.

Piketty, Thomas and Saez, Emmanuel (2007): How progressive is the US federal tax system? A historical and international perspective. *The Journal of Economic Perspectives* 21, No. 1: 3–24.

Pilgrim, Hannah, Groneweg, Merle and Reckordt, Michael (2017): The Dark Side of Digitalization: Will Industry 4.0 Create New Raw Materials Demands?. Berlin: PowerShift.

Pluta, Werner (2017): ÖPNV: Der Volocopter fliegt autonom in Dubai. *golem.de*. 27 September. Available at: www.golem.de/news/oepnv-der-volocopter-fliegt-autonom-in-dubai-1709-130297.html (accessed: 9 November 2019).

Polanyi, Karl (2010) [1944]: *The great transformation: the political and economic origins of our time*. 2. Beacon paperback ed., [reprinted]. Boston, Mass: Beacon Press.

Purdy, Mike and Paul Daugherty (2016): Why Artificial Intelligence is the future of growth. Available at: https://www.accenture.com/t20170524t055435__w__/ca-en/_acnmedia/pdf-52/accenture-why-ai-is-the-future-of-growth.pdf (accessed: 22 October 2019).

PricewaterhouseCoopers (2017): Sizing the prize. *PwC*. Available at: www.pwc.com/gx/en/issues/data-and-analytics/publications/artificial-intelligence-study.html (accessed: 22 October 2019).

PricewaterhouseCoopers (2018): Will robots really steal our jobs? An international analysis of the potential long term impact of automation. *PwC*. Available at: www.pwc.at/de/publikationen/verschiedenes/impact-of-automation-on-jobs-2018.pdf (accessed: 22 October 2019).

Privacy Rights Clearinghouse (2005/2017): Smartphone Privacy. Available at: https://privacyrights.org/consumer-guides/smartphone-privacy (accessed: 23 November 2019).

Princen, Thomas (2003): Principles for Sustainability: From Cooperation and Efficiency to Sufficiency. *Global Environmental Politics* 3, Nr. 1: 33–50.

Princen, Thomas (2005): *The logic of sufficiency*. Cambridge, MA: MIT Press.

Proske, Marina and Jaeger-Erben, Melanie (2019): Decreasing obsolescence with modular smartphones? – An interdisciplinary perspective on lifecycles. *Journal of Cleaner Production* 223 (20 June): 57–66.

Radicati Group (2019): EmailStatistics Report, 2019–2023. *Radicati*. Available at: www.radicati.com/wp/wp-content/uploads/2018/12/Email-Statistics-Report-2019-2023-Executive-Summary.pdf (accessed: 29 October 2019).

Rand Europe, RWTH Aachen and P3 Ingenieursgesellschaft (2012): Smart Trash: Study on RFID tags and the recycling industry. Study for the European Commission. Cambridge: Rand Europe.

Rankin, Jennifer (2018): Facebook, Google and Amazon could pay 'fair' tax under EU plans. *Guardian*. Available at: www.theguardian.com/business/2018/mar/21/facebook-google-and-amazon-to-pay-fair-tax-under-eu-plans (accessed: 23 October 2019).

Rauschnabel, Philipp A., Brem, Alexander and Ro, Young K. (2015): Augmented reality smart glasses: definition, conceptual insights, and managerial importance. *Unpublished Working Paper, The University of Michigan-Dearborn, College of Business*.

Raworth, Kate (2012): A safe and just space for humanity: can we live within the doughnut. *Oxfam Policy and Practice: Climate Change and Resilience* 8, No. 1: 1–26.

Reink, Michael (2016): "E-Commerce" und seine Auswirkungen auf die Stadtentwicklung. *Forum Wohnen und Stadtentwicklung* 8, No. 1.

Reisch, Lucia, Büchel, Daniela, Joost, Gesche and Zander-Hayat, Helga (2016): Digitale Welt und Handel. Verbraucher im personalisierten Online-Handel. Veröffentlichungen des Sachverständigenrats für Verbraucherfragen.

Reisch, L. A., Büchel, D., Joost, G. and Zander-Hayat, H. (2016): Consumers in the Digital World: Executive Summaries of Briefing Papers. Berlin: Sachverständigenrat für Verbraucherfragen.

Reuters (2016): Instagram's user base grows to more than 500 million. *Reuters*. Available at: www.reuters.com/article/us-facebook-instagram-users/instagrams-user-base-grows-to-more-than-500-million-idUSKCN0Z71LN (accessed: 22 October 2019).

Reuters (2019): EU fines Facebook 110 million euros over WhatsApp deal. *Reuters*. Available at: www.reuters.com/article/us-eu-facebook-antitrust/eu-fines-facebook-110-million-euros-over-whatsapp-deal-idUSKCN18E0LA (accessed: 22 November 2019).

Rifkin, Jeremy (2014): *The zero marginal cost society: The internet of things, the collaborative commons, and the eclipse of capitalism*. New York: Macmillan.

Ritch, Emma (2009): The Environmental Impact of Amazon's Kindle. San Francisco: Cleantech Group LLC.

Rogelj, Joeri, Luderer, Gunnar, Pietzcker, Robert and Riahi, Keywan (2015): Energy system transformations for limiting end-of-century warming to below 1.5°C. *Nature Climate Change* 5 (May): 519–527.

Roland Berger Strategy Consultants (2015): Die digitale Transformation der Industrie. Was sie bedeutet. Wer gewinnt. Was jetzt zu tun ist. Berlin/Munich: Bundesverband der Deutschen Industrie e.V.

Rosa, Hartmut (2003): Social acceleration: ethical and political consequences of a desynchronized high–speed society. *Constellations* 10, 3–33.

Rosa, Hartmut (2013): *Social acceleration: A new theory of modernity*. Columbia University Press.

Rösch, C., Dusseldorp, M. and Meyer, R. (2005): Precision Agriculture. *Office of technology assessment at the German Bundestag*, Working report no. 106. Available at: www.tab-beim-bundestag.de/en/pdf/publications/summarys/TAB-Arbeitsbericht-ab106_Z.pdf (accessed: 19 November 2019).

Ross, B., Pilz, L., Cabrera, B., Brachten, F., Neubaum, G. and Stieglitz, S. (2019): Are social bots a real threat? An agent-based model of the spiral of silence to analyse the impact of manipulative actors in social networks. European *Journal of Information Systems*.

Roser, Max and Ritchie, Hannah (2017): Technological Progress. *Our World in Data*. Available at: https://ourworldindata.org/technological-progress (accessed: 6 December 2019).

Rosqvist, Lena Smidfelt and Hiselius, Lena Winslott (2016): Online shopping habits and the potential for reductions in carbon dioxide emissions from passenger transport. *Journal of Cleaner Production* 131 (10 September): 163–169.

Rothstein, Edward (2006): A Crunchy-Granola Path From Macramé and LSD to Wikipedia and Google. *New York Times*. 25 September. Available at: www.nytimes.com/2006/09/25/arts/25conn.html?_r=0 (accessed: 24 October 2019).

Rubin, Julia, Gordon, Michael I., Nguyen, Nguyen and Rinard, Martin (2015): Covert Communication in Mobile Applications. *MIT web domain* (November). Available at: http://dspace.mit.edu/handle/1721.1/99941 (accessed: 6 December 2019).

Rushkoff, Douglas (2016): *Throwing rocks at the Google bus: how growth became the enemy of prosperity*. New York: Portfolio Trade.

Rüßmann, Michael, Lorenz, Markus, Gerbert, Philipp, Waldner, Manuela, Justus, Jan, Engel, Pascal and Harnisch, Michael (2015): Industry 4.0: The Future of Productivity and Growth in Manufacturing Industries. Published by Boston Consulting Group.

Sachs, Jeffrey D., Modi, Vijay, Figueroa, Hernan, Machado Fantacchiotti, Mariela, Sanyal, Kayhan, Khatun, Fahmida and Shah, Aditi (2015): ICT & SDGs. How Information and Communications Technology Can Achieve The Sustainable Development Goals. Columbia University.

Sachs, Jeffrey D. (2018): America's Health Crisis and the Easterlin Paradox. *World Happiness Report 2018*, pp. 146–159. Available at: https://s3.amazonaws.com/happiness-report/2018/CH7-WHR-lr.pdf (accessed: 6 December).

Sachs, Wolfang and Santarius, Tilman (2007): *Slow Trade-Sound Farming A Multilateral Framework for Sustainable Markets in Agriculture*. Berlin/Aachen: Heinrich-Böll-Foundation/Misereor.

Sachs, Wolfgang (1984): *Die Liebe zum Automobil. Ein Rückblick in die Geschichte unserer Wünsche*. Reinbek: Rowohlt.

Sachs, Wolfgang (1993): Die vier E's: Merkposten für einen maß-vollen Wirtschaftsstil. *Politische Ökologie* 11, No. 33: 69–72.

Sachs, Wolfgang (2015): *Planet dialectics: explorations in environment and development. 2nd Edition.* London: Zed books.

Sachs, Wolfgang and Santarius, Tilman, Eds (2007): *Fair Future. Limited Resources and Global Justice.* London: Zed books.

Sachverständigenrat für Verbraucherfragen (2016): Verbraucherrecht 2.0 Verbraucher in der digitalen Welt. Berlin: Sachverständigenrat für Verbraucherfragen.

Sacom (2010): Workers as Machines: Military Management in Foxconn. Students & Scholars Against Corporate Misbehaviour.

Salahuddin, Mohammad and Alam, Khorshed (2016): Information and Communication Technology, electricity consumption and economic growth in OECD countries: A panel data analysis. *International Journal of Electrical Power & Energy Systems* 76 (March): 185–193.

Saner, R., Yiu, L. and Nguyen, M. (2018): Platform Cooperatives: The Social and Solidarity Economy and the Future of Work: A Preliminary Assessment of Platform Capitalism and Platform Cooperativism and their Effects on Workers' Satisfaction. *The UN Inter-Agency Task Force on Social and Solidarity Economy (TFSSE).* Available at: https://unsse.org/wp-content/uploads/2019/06/Saner_Platform-Cooperatives_En.pdf (accessed: 25 November 2019).

Santarius, Tilman (2016a): Investigating meso-economic rebound effects: production-side effects and feedback loops between the micro and macro level. *Journal of Cleaner Production* 134 (October): 406–413.

Santarius, Tilman (2016b): Energy Efficiency and Social Acceleration: Macro-level Rebounds from a Sociological Perspective. *Rethinking Climate and Energy Policies. New Perspectives on the Rebound Phenomenon.* Springer, New York, 143–161.

Santarius, Tilman, Walnum, Hans Jakob and Aall, Carlo (2016): *Rethinking Climate and Energy Policies: New Perspectives on the Rebound Phenomenon.* Cham: Springer.

Satariano, Adam, Russell, Karl, Griggs, Troy, Migliozzi, Blacki and Lee, Chang W. (2019): How the Internet Travels Across Oceans. *New York Times* (10 March) Available at: www.nytimes.com/interactive/2019/03/10/technology/internet-cables-oceans.html (accessed: 6 December 2019).

Saunders, Harry D. (2013): Historical evidence for energy efficiency rebound in 30 US sectors and a toolkit for rebound analysts. *Technological Forecasting and Social Change* 80, No. 7: 1317–1330.

Schaar, Peter (2010): Privacy by Design. *Identity in the Information Society* 3, Nr. 2 (August): 267–274.

Schien, Daniel, Shabajee, Paul, Yearworth, Mike and Preist, Chris (2013): Modeling and Assessing Variability in Energy Consumption During the Use Stage of Online Multimedia Services: Energy Consumption During Use of Online Multimedia Services. *Journal of Industrial Ecology* 17, No. 6 (December): 800–813.

Schiller, Dan (2000): *Digital capitalism: networking the global market system.* 1. paperback. Cambridge, Massachusetts, London: MIT Press.

Schneidewind, Uwe, Santarius, Tilman and Humburg, Anja, eds. (2013): *Economy of sufficiency: essays on wealth in diversity, enjoyable limits and creating commons.* Wuppertal spezial 48. Wuppertal: Wuppertal Inst. for Climate, Environment and Energy.

Scholz, Trebor (2016a): Platform Cooperativism. Challenging the Corporate Sharing Economy. New York: Rosa Luxemburg Stiftung.

Scholz, Trebor (2016b): *Uberworked and underpaid: how workers are disrupting the digital economy.* Hoboken, New Jersey: John Wiley & Sons.

Schor, Juliet B. (1998): *The Overspent American. Why we want what we don't need.* New York: Basic Books.

Schor, J. B. and Attwood-Charles, W. (2017): The Bsharing^economy: labor, inequality, and social connection on for-profit platforms. Sociology Compass, 11(8), 1–16.

Schulze, G. (2013): The experience market. Sundbo J, Sørensen F(eds): *Handbook on the experience economy.* Edward Elgar Publishing, pp. 98–122.

Schumacher, E.F. (1973): *Small is beautiful. A study of economics as if people mattered.* London: Blond & Briggs.

Schumpeter, Joseph Alois (1942): *Capitalism, socialism and democracy.* New York, London: Harper.

Schwarz, Hunter (2012): How Many Photos Have Been Taken Ever? *Buzzfeed.* 24 September. Available at: www.buzzfeed.com/hunterschwarz/how-many-photos-have-been-taken-ever-6zgv?utm_term=.irrDP884o#.aeWW9OOGR (accessed: 28 June 2019).

Seetharam, Anand, Somasundaram, Manikandan, Towsley, Don, Kurose, Jim and Shenoy, Prashant (2010): Shipping to Streaming: Is this shift green? *Proceedings of the first ACM SIGCOMM workshop on Green networking,* 61–68.

Sekulova, Filka and Schneider, Francois (2014): Open Localism. Event: Degrowth Conference Leipzig 2014, Leipzig.

Shaheen, S., Cohen, A. and Martin, E. (2017): Smartphone App Evolution and Early Understanding from a Multimodal App User Survey. UC Berkeley. DOI10.7922/G2CZ35CH.

Shane, S., Rosenberg, M. and Lehren, A. W. (2017): WikiLeaks Releases Trove of Alleged C.I.A. Hacking Documents. *New York Times* www.nytimes.com/2017/03/07/world/europe/wikileaks-cia-hacking.html (accessed: 4 November 2019).

Shearer, Chad, Vey, Jennifer S. and Kim, Joanne (2019): Where jobs are concentrating and why it matters to cities and regions. Brookings. Available at: www.brookings.edu/research/where-jobs-are-concentrating-why-it-matters-to-cities-and-regions/(accessed: 28 October 2019).

Shehabi, Arman, Walker, Ben and Masanet, Eric (2014): The energy and greenhouse-gas implications of internet video streaming in the United States. *Environmental Research Letters* 9, No. 5 (1 May): 054007.

Siikavirta, Hanne, Punakivi, Mikko, Kärkkäinen, Mikko and Linnanen, Lassi (2002): Effects of E-Commerce on Greenhouse Gas Emissions: A Case Study of Grocery Home Delivery in Finland. *Journal of Industrial Ecology* 6, No. 2 (1 April): 83–97.

Silver, David, Schrittwieser, Julian, Simonyan, Karen, Antonoglou, Ioannis, Huang, Aja, Guez, Arthur, Hubert, Thomas, Baker, Lucas, Lai, Matthew, Bolton, Adrian, *et al.* (2017): Mastering the game of Go without human knowledge. *Nature* 550, No. 7676 (19 October): 354–359.

Sivaraman, Deepak, Pacca, Sergio, Mueller, Kimberly and Lin, Jessica (2007): Comparative energy, environmental, and economic analysis of traditional and e-commerce DVD rental networks. *Journal of Industrial Ecology* 11, No. 3: 77–91.

Skerrett, Ian (2017): IoT Developer Survey 2017. *Slideshare.* Available at: www.slideshare.net/IanSkerrett/iot-developer-survey-2017 (accessed: 24 November 2019).

Skjelvik, J. M., Erlandsen, A. M. and Haavardsholm, O. (2017): Environmental impacts and potential of the sharing economy. Available at: http://dx.doi.org/10.6027/TN2017-554.

Smith, A., Fressoli, M., Abrol, D., Arond, E. and Ely, A. (2016): Grassroots Innovation Movements. Routledge.

Social Innovation Community (2018): What is Social Innovation?. *SIC.* Available at: www.siceurope.eu/about-sic/what-socil-innovation/what-social-innovation (accessed: 27 November 2019).

Sokolov, Daniel A. J. (2016): Forscher: Selbstfahrende Autos bringen Verkehrslawine. *Heise online*. Available at: www.heise.de/newsticker/meldung/Forscher-Selbstfahrende-Autos-bringen-Verkehrslawine-3070571.html (accessed: 1 August 2019).

Solon, Olivia (2017): Is Lyft really the "woke" alternative to Uber? *Guardian*. 29 March. Available at: www.theguardian.com/technology/2017/mar/29/is-lyft-really-the-woke-alternative-to-uber (accessed: 16 October 2019).

Song, Indeok, Larose, Robert, Eastin, Matthew and Lin, Carolyn (2004): Internet Gratifications and Internet Addiction: On the Uses and Abuses of New Media. Cyberpsychology & behavior: the impact of the Internet, multimedia and virtual reality on behavior and society. DOI:7. 384–94. 10.1089/cpb.2004.7.384.

Sorrell, Steve (2009): Jevons' Paradox revisited: The evidence for backfire from improved energy efficiency. *Energy Policy* 37, No. 4 (April): 1456–1469.

Staab, Philipp (2017): The consumption dilemma of digital capitalism. *Transfer: European Review of Labour and Research, 23*(3), 281–294.

Staley, Oliver (2018): There's a new list of the world's 10 largest companies—and tech isn't on it. *Quartz*. Available at: https://qz.com/1331995/walmart-is-the-worlds-biggest-company-apple-isnt-in-the-top-10/ (accessed: 21 November 2019).

Stallman, Richard (1985): *The GNU manifesto.*

Standing, Guy (2018): The Precariat: Today's Transformative Class? *Great Transition Initiative*. Available at: www.greattransition.org/publication/precariat-transformative-class. (accessed: 18 October 2019).

Stanford Graduate School of Business (2019): Defining Social Innovation. *Stanford Graduate School of Business*. Available at: www.gsb.stanford.edu/faculty-research/centers-initiatives/csi/defining-social-innovation (accessed: 27 November 2019).

StatCounter (2019a): Desktop Operating System Market Share Worldwide. Available at: https://gs.statcounter.com/os-market-share/desktop/worldwide/ (accessed: 19 November 2019).

StatCounter (2019b): Global market share held by leading desktop internet browsers from January 2015 to June 2019. *Statista*. Available at: www.statista.com/statistics/544400/market-share-of-internet-browsers-desktop/ (accessed: 16 October 2019).

StatCounter (2019c): Mobile Operating System Market Share Worldwide. Available at: https://gs.statcounter.com/os-market-share/mobile/worldwide (accessed: 19 November 2019).

StatCounter (2019d): Search Engine Market Share Worldwide. Available at: https://gs.statcounter.com/search-engine-market-share (accessed: 22 November 2019).

StatCounter (2019e): Social Media Stats Worldwide. Available at: https://gs.statcounter.com/social-media-stats (accessed: 19 November 2019).

Statista (2019a): Amazon Challenges Ad Duopoly. Available at: www.statista.com/chart/17109/us-digital-advertising-market-share/ (accessed: 22 November 2019).

Statista (2019b): Most popular social networks worldwide as of October 2019, ranked by number of active users. Available at: www.statista.com/statistics/272014/global-social-networks-ranked-by-number-of-users/ (accessed: 22 November 2019).

Statistisches Bundesamt (2013): Broschüre Verkehr auf einen Blick. Wiesbaden.

Statistisches Bundesamt (2017): Verkehr aktuell. Fachserie 8 Reihe 1.1. Wiesbaden.

Steffen, Will, Richardson, Katherine, Rockström, Johan, Cornell, Sarah E., Fetzer, Ingo, Bennett, Elena M., Biggs, Reinette, Carpenter, Stephen R., Vries, Wim de, Cynthia, de Wit, A. *et al.* (2015): Planetary boundaries: Guiding human development on a changing planet. *Science* 347, No. 6223.

Steinberg, Philip E., Nyman, Elizabeth and Caraccioli, Mauro J. (2012): Atlas Swam: Freedom, Capital, and Floating Sovereignties in the Seasteading Vision. *Antipode* 44, No. 4 (September): 1532–1550.

Sternberg, André and Bardow, André (2015): Power-to-What? – Environmental assessment of energy storage systems. *Energy & Environmental Science* 2, 2015 (20 January): 389–400.

Stobbe, Lutz, Proske, Marina, Zedel, Hannes, Hintemann, Ralph, Clausen, Jens and Beucker, Severin (2015): Entwicklung des IKT-bedingten Strombedarfs in Deutschland. Abschlussbericht. Berlin: Fraunhofer-Institut für Zuverlässigkeit und Mikrointegration (IZM).

Stockhammer, Engelbert (2017): Determinants of the Wage Share: A Panel Analysis of Advanced and Developing Economies: Determinants of the *Wage Share*. *British Journal of Industrial Relations* 55, No. 1 (March): 3–33.

Strubell, Emma, Ganesh, Ananya and McCallum, Andrew (2019): Energy and Policy Considerations for Deep Learning in NLP. *arXiv:1906.02243 [cs]* (5 June).

Sullivan, Danny (2015): How Machine Learning Works, As Explained By Google. *MarTech Today*. 4 November. Available at: https://martechtoday.com/how-machine-learning-works-150366 (accessed: 9 November 2019).

Summers, Larry (2013): Larry Summers at IMF Economic Forum. Vortrag auf der 14th Annual IMF Research Conference: Crises Yesterday and Today, 8 November. *YouTube*. Available at: www.youtube.com/watch?v=KYpVzBbQIX0 (accessed: 6 December 2019).

Tapscott, Don and Tapscott, Alex (2016): *Blockchain revolution: the brilliant technology changing money, business and the world*. New York: PenguinRandomHouse.

Ternes, Anabel, Towers, Ian and Jerusel, Marc (2015): *Konsumentenverhalten im Zeitalter der Digitalisierung*. Wiesbaden: Springer Fachmedien.

Teulings, Coen and Baldwin, Richard, Eds (2014): *Secular Stagnation: Facts, Causes and Cures*. London: CEPR Press.

The Economist (2012): The last Kodak moment? *The Economist*. Available at: www.economist.com/node/21542796 (accessed: 27 October 2017).

The Economist (2016): Android attack. *The Economist* (23 April). Available at: www.economist.com/news/business/21697193-european-commission-going-after-google-againthis-time-better-chance (accessed: 28 October 2019).

The Economist (2017): Regulating the internet giants: The world's most valuable resource is no longer oil, but data. *The Economist*. Available at: www.economist.com/leaders/2017/05/06/the-worlds-most-valuable-resource-is-no-longer-oil-but-data (accessed: 23 November 2019).

The Economist Data Team (2014): "Secular stagnation" in graphics – Doom and gloom. *The Economist* (20 November) Available at: www.economist.com/blogs/graphicdetail/2014/11/secular-stagnation-graphics (accessed: 14 October 2019).

The Green Press Initiative (2011): Environmental Impacts of E-Books. *The Green Press Initiative*. Available at: www.greenpressinitiative.org/documents/ebooks.pdf.

The New Climate Institute (2016): What does the Paris Agreement mean for climate protection in Germany? Berlin: Greenpeace. Available at: https://newclimate.org/2016/02/23/what-does-the-paris-agreement-mean-for-climate-protection-in-germany/ (accessed: 2 December 2019).

The Shift Project (2019): Lean ICT: Towards digital sobriety. Paris.

The World Development Report (WDR) (2019): The Changing Nature of Work studies. *The World Bank*. Available at: www.worldbank.org/en/publication/wdr2019 (accessed: 22 October 2019).

ThingsCon (2017): The State of Responsible IoT. Berlin.

Toffler, Alvin (1980): *The Third Wave*. New York: Bantam Books.

Tracxn (2019): Top Multimodal Transport Apps Startups. *Tracxn*. Available at: https://tracxn.com/d/trending-themes/Startups-in-Multimodal-Transport-Apps/ (accessed: 12 November 2019).

Trapp, Katharina (2015): Measuring the labour income share of developing countries. WIDER Working Paper 2015/041. World Institute for Development Economics Research.

Trommer, Stefan, Kolarova, Viktoriya, Fraedrich, Eva, Kröger, Lars, Kickhöfer, Benjamin, Kuhnimhof, Tobias, Lenz, Barbara and Phleps, Peter (2016): Autonomous Driving: The Impact of Vehicle Automation on Mobility Behaviour. Berlin: DLR & ifmo.

Türk, Volker, Alakeson, Vidhja, Kuhndt, Michael and Ritthoff, Mmichael (2003): The environmental and social impacts of digital music. A case study with EMI. Wuppertal.

Turner, Fred (2005): Where the counterculture met the new economy: The WELL and the origins of virtual community. *Technology and Culture* 46, No. 3: 485–512.

Turner, Fred (2010): *From counterculture to cyberculture: Stewart Brand, the Whole Earth Network, and the rise of digital utopianism*. Chicago and London: The University of Chicago Press.

Turner, Karen (2009): Negative rebound and disinvestment effects in response to an improvement in energy efficiency in the UK economy. *Energy Economics* 31, No. 5: 648–666.

UNFCCC (2015): Paris Agreement. www.unfccc.int.

United Nations (1966): *International Covenant on Civil and Political Rights*.

United Nations (2015): *Transforming our world: the 2030 Agenda for Sustainable Development*.

Van den Boer, Arnould (2015): Dynamic pricing and learning: historical origins, current research, and new directions. Surveys in operations research and management science, 20(1): 1–18.

Vardi, Moshe (2017): Humans, Machines, and Work: The Future is Now. Vortrag. Event: Technology, Cognition and Culture Lecture Series, 27 March, Ken Kennedy Institute for Information Technology, Houston. www.youtube.com/watch?v=5ThiClGEBes.

Verband privater Rundfunk und Telemedien (2016): Grafiken zur Mediennutzungsanalyse 2016. *Vaunet*. Available at: www.vau.net/pressebilder/content/grafiken-mediennutzungsanalyse-20161 (accessed: 10 December 2019).

Vermesan, Ovidiu and Friess, Peter (2016): *Digitising the Industry: Internet of Things Connecting the Physical, Digital and Virtual Words*. River Publishers Series in Communications 49. Aalborg: River Publishers.

Vezzoli, Carlo, Ceschin, Fabrizio, Osanjo, Lilac, M'Rithaa, Mugendi K., Moalosi, Richie, Nakazibwe, Venny and Diehl, Jan Carel (2018): *Designing Sustainable Energy for All*. Green Energy and Technology. Cham: Springer International Publishing.

Vezzoli, Carlo, Ceschin, Fabrizio, Osanjo, Lilac, M'Rithaa, Mugendi K., Moalosi, Richie, Nakazibwe, Venny and Diehl, Jan Carel (2018): Distributed/Decentralised Renewable Energy Systems. *Designing Sustainable Energy for All*, 23–39. Cham: Springer International Publishing.

Victor, Peter A. (2019): *Managing without growth: slower by design, not disaster, Second Edition*. Cheltenham: Edward Elgar Publishing.

Voshmgir, Shermin (2016): Blockchains, Smart Contracts und das Dezentrale Web. Berlin: Technologiestiftung Berlin. Available at: www.technologiestiftung-berlin.de/fileadmin/daten/media/publikationen/170130_BlockchainStudie.pdf.

Wachs, Audrey (2017): Seasteaders to bring a libertarian floating community to the South Pacific. *Archpaper.com*. Available at: https://archpaper.com/2017/10/seasteading-institute-floating-libertarian-ocean-cities/ (accessed: 2 November 2019).

Wahnbaeck, Carolin and Roloff, Lu Yen (2017): After the Binge, the Hangover. Insights into the Minds of Clothing Consumers. Hamburg: Greenpeace.

Wakefield, Jane (2019): Christchurch shootings: Social media races to stop attack foot-age. *BBC News.* Available at: www.bbc.com/news/technology-47583393 (accessed: 22 November 2019).

Walker, Alissa (2014): A Map of Every Device in the World That's Connected to the Inter-net. *Gizmodo.* Available at: https://gizmodo.com/a-map-of-every-device-in-the-world-thats-connected-to-t-1628171291 (accessed: 12 October 2019).

Wang, Rebecca Jen-Hui, Malthouse ,Edward C. and Krishnamurthi, Lakshman (2015): On the Go: How Mobile Shopping Affects Customer Purchase Behavior. *RETAIL Journal of Retailing* 91, No. 2 (June): 217–234.

Wang, Y. and Hao, F. (2018): Does Internet penetration encourage sustainable consump-tion? A cross-national analysis. Sustainable Production and Consumption, 16, 237–248.

Weber, Christopher L., Koomey, Jonathan G. and Scott Matthews, H. (2010): The Energy and Climate Change Implications of Different Music Delivery Methods. *Journal of Indus-trial Ecology* 14, No. 5 (October): 754–769.

Weber, C., Koyama, M. and Kraines, S. (2006): CO_2-emissions reduction potential and costs of a decentralized energy system for providing electricity, cooling and heating in an office-building in Tokyo. *Energy* 31, Nr. 14 (November): 3041–3061.

Weber, K., Matthias, Gudowsky, Niklas, Aichholzer, Georg (2018): Foresight and technol-ogy assessment for the Austrian parliament — Finding new ways of debating the future of industry 4.0. *Futures*, Volume 109, 240–251.

Weizsäcker, Ernst Ulrich, Hargroves, Karlson, Smith, Michael, Desha, Cheryl and Stasinopoulos, Peter (2009): *Factor 5: Transforming the Global Economy.* London: Earthscan.

Welzer, Harald (2016): Smart Dictatorship: The Assault on Our Freedom. S. Fischer Verlag GmbH, Frankfurt am Main.

Wibe, Sören and Carlén, Ola (2006): Is Post-War Economic Growth Exponential? *Australian Economic Review* 39, No. 2 (1 June): 147–156.

Wiese, Anne, Toporowski, Waldemar and Zielke, Stephan (2012): Transport-related CO_2 effects of online and brick-and-mortar shopping: A comparison and sensitivity analysis of clothing retailing. *Transportation Research Part D: Transport and Environment* 17, No. 6 (August): 473–477.

Wikipedia.org, Ed. (2017a): Wikipedia. *Wikipedia.* 23 June. https://de.wikipedia.org/w/index.php?title=Wikipedia&oldid=166640826.

Wikipedia.org, Ed. (2017b): Distributed economy. *Wikipedia.* 11 September. Available at: https://en.wikipedia.org/w/index.php?title=Distributed_economy&oldid=800083010 (accessed: 6 December 2019).

Wilkinson, Richard G. and Pickett, Kate (2010): *The Spirit Level: Why Equality is better for Everyone.* Penguin sociology. London: Penguin Books.

Wilson, Mark (2017): The Web Is Basically One Giant Targeted Ad Now. *Co.Design.* 1 June. Available at: www.fastcodesign.com/90127650/the-web-is-basically-one-giant-targeted-ad-now (accessed: 28 June 2019).

Winkel, Olaf (2016): The Perspectives of Democratic Decision Making in the Information Society. *IJCSIT – International Journal of Computer Science and Information Technology,* No. 2/2016: 101–116.

Witte, Klemens (2019): The effects of automation on unemployment and mental health: Is universal basic income the ultimate cure? *DOC Research Institute.* Available at: https://doc-research.org/2019/07/automation-unemployment-mental-health-ubi/ (accessed: 11 October 2019).

Wittbrodt, B.T., Glover, A.G., Laureto, J., Anzalone, G.C., Oppliger, D., Irwin, J. and L. and Pearce, J.M. (2013): Life-cycle economic analysis of distributed manufacturing with open-source 3-D printers. *Mechatronics* 23, No. 6: 713–726.

Wong, Julia Carrie (2017): Seasteading: tech leaders' plans for floating city trouble French Polynesians. *Guardian* (2 January), Technology section. www.theguardian.com/technology/2017/jan/02/seasteading-peter-thiel-french-polynesia.

World Economic Forum (2013): Breaking the Binary: Policy Guide to Scaling Social Innovation. http://reports.weforum.org/social-innovation-2013/ (accessed: 25 November 2019).

World Economic Forum (2018a): 5 things to know about the future of jobs. *World Economic Forum*. Available at: www.weforum.org/agenda/2018/09/future-of-jobs-2018-things-to-know/ (accessed: 19 November 2019).

World Economic Forum (2018b): Future of Jobs Report 2018. *World Economic Forum*. Available at: www3.weforum.org/docs/WEF_Future_of_Jobs_2018.pdf (accessed: 18 October 2019).

World Economic Forum (2019): With lifelong learning, you too can join the digital workplace. *World Economic Forum*. Available at: www.weforum.org/agenda/2019/08/lifelong-learning-in-the-digital-workplace-is-essential-heres-why/ (accessed: 18 October 2019).

World Energy Council (2016): The road to resilience: Managing cyber risks. *Worldenergy.org*. September. Available at: www.worldenergy.org/publications/2016/the-road-to-resilience-managing-cyber-risks/.

World Inequality Lab (2018): World Inequality Report, Executive Summary. Available at: https://wir2018.wid.world/files/download/wir2018-summary-english.pdf (accessed: 27 November 2019).

Wu, Tim (2002): A Proposal for Network Neutrality. Charlottesville.

Xu, Zheng (2019): The electricity market design for decentralized flexibility sources. Oxford: Oxford Institute for Energy Studies.

Zago, Matteo Gianpietro (2018): Why the Web 3.0 Matters and you should know about it. *Medium*. Available at: https://medium.com/@matteozago/why-the-web-3-0-matters-and-you-should-know-about-it-a5851d63c949. (accessed: 23 October 2019).

Zierahn, Ulrich, Gregory, Terry and Arntz, Melanie (2016a): Racing With or Against the Machine? Evidence from Europe. Discussion Paper No. 16–053, ZEW Centre for European Economic Research.

Zierahn, Ulrich, Gregory, Terry and Arntz, Melanie (2016b): The risk of automation for jobs in OECD countries: a comparative analysis, OECD Social, Employment and Migration Working Papers No 189.

Zimmermann, Hendrik and Wolf, Verena (2016): Sechs Thesen zur Digitalisierung der Energiewende: Chancen, Risiken und Entwicklungen. Bonn, Berlin: Germanwatch e.V.

Zuboff, Shoshana (2019): *The age of surveillance capitalism: the fight for a human future at the new frontier of power*. New York: PublicAffairs.

Zuo, Ying, Tao, Fei and Nee, A.Y. C. (2017): An Internet of things and cloud-based approach for energy consumption evaluation and analysis for a product. *International Journal of Computer Integrated Manufacturing* (6 February): 1–12.

INDEX

Page numbers in **bold** denote tables, those in *italics* denote figures.